THE
CABARET
OF
PLANTS

THE
CABARET
OF
PLANTS

FORTY THOUSAND YEARS OF PLANT LIFE AND THE HUMAN IMAGINATION

RICHARD MABEY

W. W. NORTON & COMPANY
Independent Publishers Since 1923
New York · London

For information about permission to reproduce selections from
this book, write to Permissions, W. W. Norton & Company, Inc.,
500 Fifth Avenue, New York, NY 10110

For information about special discounts for bulkpurchases, please contact
W. W. Norton Special Sales at specialsales@wwnorton.com or 800-233-4830

Manufacturing by Versa Press
Production manager: Julia Druskin

Library of Congress Cataloging-in-Publication Data

Mabey, Richard, 1941– author.
The cabaret of plants : forty thousand years of plant life and the human
imagination / Richard Mabey.—First American edition.
pages cm
First published in Great Britain in 2015 under the title:
The cabaret of plants: botany and the imagination.
Includes bibliographical references and index.
ISBN 978-0-393-23997-3 (hardcover)
1. Botany—History. I. Title.
QK15.M32 2016
580—dc23
2015033568

W. W. Norton & Company, Inc.
500 Fifth Avenue, New York, N.Y. 10110
www.wwnorton.com

W. W. Norton & Company Ltd.
Castle House, 75/76 Wells Street, London W1T 3QT

1 2 3 4 5 6 7 8 9 0

for Vivien

The force that through the green fuse drives the flower
Drives my green age …

Dylan Thomas

Contents

Introduction: The Vegetable Plot

JUST BEFORE HE DIED in 1888 Edward Lear sketched the last of the surreal additions to evolution's menagerie that he'd begun with the Bong-tree in 'The Owl and the Pussycat' nearly twenty years before. His *Nonsense Botany* is a series of impish cartoons of preposterous floral inventions. It includes a strawberry bush bearing puddings instead of fruit, the parrot-flowered *Cockatooca superba* and the unforgettable *Manypeeplia upsidownia*, a kind of Solomon's seal with minute humans suspended like flowers along the bowed stalk. Lear was a lifelong sufferer from epilepsy and depressive episodes (which he nicknamed 'the Morbids' as if they were a tribe of gloomy rodents) and the obsessive fun he had with words and forms may have been a way of exorcising his melancholy. But I suspect there is more to his final creation. Lear was an astute botanist as well as a brilliant humorist. He'd travelled and painted across the Old World, especially in the Mediterranean, and had seen first hand many of its bizarre plants, including the carrion-stinking dragon arum (which he described as 'brutal-filthy yet picturesque'), and I think his nonsense flora can be seen as a kind of celebratory cabaret, an affectionate satire on the astonishing revelations of nineteenth-century botany.

Thirty years previously Europeans had their first news of the Welwitschia, a Namibian desert plant whose single pair of leaves can live for 2,000 years, grow to immense size but remain in the permanently infantilised state of a seedling. Ten years later Charles Darwin had revealed the barely credible devices orchids used to conscript

1

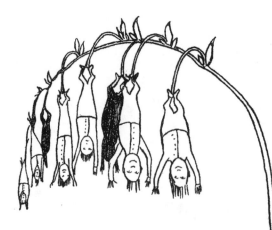

Manypeeplia upsidownia,
from Edward Lear's *Nonsense Botany.*

insect pollinators, including the launching of pollen-laden missiles. In a world of such remarkable organisms why shouldn't there be a fly orchid dangling real flies like Lear's *Bluebottlia buzztilentia*? As for his *Sophtsluggia glutinosa*, it could well be the filthy dragon arum reimagined as one of the plant–animal cooperatives being unmasked by explorers in the tropics. Lear's bionic vegetables were botany's reductio ad absurdum, the last tarantellas of a century in which plants had been just about the most interesting things on the planet. It wasn't a fascination confined to the scientific elite. The general public had been agog, astounded by one botanical revelation after another. In America the discovery of the ancient sequoias of California in the 1850s drew tens of thousands of pilgrims, who saw in these giant veterans proof of their country's manifest destiny as an unsullied Eden. (There were throngs of rubberneckers and partygoers too: nineteenth-century botany was far from sober-sided.) Similar numbers flocked to Kew Gardens in west London, where one of the star attractions was an Amazonian water lily whose leaves were so brilliantly engineered that their design became the model for the greatest glass building of the nineteenth century. What these moments of excited attention shared was not so much a simple pleasure in floral beauty or the promise of new sources of imperial revenue (though these were there too) but a sense of real wonder that units of non-conscious green tissue could have such strange existences and unquantifiable powers. Plants, defined by their immobility, had evolved extraordinary life-ways by way of compensation: the power to regenerate after most of their body had been eaten; the ability to have sex by proxy; the possession of more than twenty senses whose delicacy far exceeded any of our own. They made you *think*.

Yet if respect for them as complex and adventurous organisms reached its zenith in the late nineteenth century, it neither began nor ended there. People had been enthralled by and sometimes fearful of the vegetable world's alternative solutions to living for thousands of years. They contrived myths to explain why trees could outlive civilisations; invented hybrid creatures – chimera – as models for plants they were unable to understand, and which seemed to intuit symbioses discovered centuries later. Ironically, the same scientific revolution that engaged the public imagination eventually alienated it. Alongside Darwin's work, Gregor Mendel's discoveries of the mechanisms of genetic inheritance in the late 1860s drove botany deeper in to the laboratory. The workings of plants became too difficult, too intricate for popular understanding. Amateur botanists turned instead to recording the distribution of wild species. The rest of us mostly sublimated our interest in the existence of plants into pleasure at their outward appearance, and the garden has become the principal theatre of vegetal appreciation. Plants in the twenty-first century have been largely reduced to the status of utilitarian and decorative objects. They don't provoke the curiosity shown to, say, dolphins or birds of prey or tigers – the charismatic celebrities of television shows and conservation campaigns. We tend not to ask questions about how they behave, cope with life's challenges, communicate both with each other and, metaphorically, with us. They have come to be seen as the furniture of the planet, necessary, useful, attractive, but 'just there', passively vegetating. They are certainly not regarded as 'beings' in the sense that animals are.

This book is a challenge to that view. It's a story about plants as authors of their own lives and an argument that ignoring their vitality impoverishes our imaginations and our well-being. It begins with the very first representations of plants in cave art 35,000 years ago, and the revelation that Palaeolithic artists were more intrigued by plants as forms than food. And it ends in a kind of modern cave: the hollow shell of a famous fallen beech, and what this apparently dead relic says about the ability of plants, working as a community, to survive catastrophe. In between, I discuss how medieval clerics and indigenous shamans laid down formal explanations of why one wilding could evolve into a food crop and another into a poison; the debate between Romantic poets and Enlightenment scientists about the kind of vital forces that might lie behind vegetal powers, and whether they echoed the creativity of humans; and today, how the puzzles that so excited the nineteenth century – do plants have intentions? inventiveness? individuality? – are being explored by a new breed of unconventional and multidisciplinary thinkers.

It's odd that we haven't regained our ancient sense of wonder, espe-cially now we understand how crucial the plant world is to our own survival. Perhaps that is partly the answer: we find it hard to accept that plants don't need us in the way we need them. The UN has described the 300,000-plus species which make up the earth's flora as 'the economy's primary producer ... photosynthetic cells capture a proportion of the sun's radiant energy and from that silent diurnal act comes everything we have: air to breathe, water to drink, food to eat, fibres to wear, medicines to take, timber for shelter'. They are now a front-line crisis service too. Trees combat climate change, soak up floods, purify city air. Wild flowers help insects survive so they can pollinate human crops. The structures of plant tissues are providing models for a new genera-tion of engineered, non-polluting materials. You would think that this increasing understanding of the centrality of plants' role on earth might encourage a new respect for them as autonomous organisms. But the opposite is happening. Influential conservationists such as Tony Juniper have openly abandoned the idea of arguing for plants' 'intrinsic value' in favour of stressing their economic potential, and have enthusiastically embraced the jargon of the marketplace. Wordsworth's 'host of golden daffodils' has been rebranded as 'natural capital' and the Wildwood as a provider of 'ecosystem services'. 'Nature', once seen as some kind of alternative or counter to the ugliness of corporate existence, is now being sucked into it. I've no doubt that the pragmatic realpolitik and self-interest of this approach are powerful motivators for conservation. But I think of George Orwell's words: 'if thought corrupts language, language can also corrupt thought'. And worry about the subliminal effects of defining plants as a biological proletariat, working solely for the benefit of our species, without granting them any a priori importance. One doesn't have to believe that plants have rights to see that this is a precarious status, subject to the swings of human taste and attitude. In the absence of respect and real curiosity, attentiveness falters. Complex systems become reduced to green blurs, with dangerous consequences both for us and for individual species. An example of well-meaning but myopic human-centred thinking is the encouragement being given to the growing of nectar-rich flowers for pollinators, bees especially. This is a commendable policy – except that the majority of pollinating insects, unlike bees, grow from larvae that feed not on nectarous flowers but dull green leaves, some of them the weeds that are hoicked out to make way for the dazzling floral border.

I suspect that the chief problem we have in considering plants as autonomous beings – and let me risk embarrassment by using the word *selves* here, meaning authors of their own life stories – is that they seem to have no animating spirit. I was lucky in that I had an early and revelatory glimpse of their vitality. I tell the full story later of how one species, marsh samphire, took me through this transformative experience, but its outlines are relevant here. I first came across the plant as a commodity, a foodstuff, a very desirable wild delicacy (I still rate it foraging's gold star species), and then discovered that it had an enthralling existence beyond my use for it – a love for bare, viscous mudflats which was seemingly contradicted by an inherent drive to turn them into dry land.

Most of my personal encounters with plants – some of which are described in the pages that follow – have confirmed this conviction that plants have agendas of their own. On every occasion I have owned, or had control of, or planned purposes for vegetation, what has enthralled me has been the way the plants go off on courses entirely of their own. During the years I owned the deeds of an ancient wood in the Chilterns (I can't say I truly owned the living place itself) our feeble attempts at tree planting were swamped by the wood's decision to grow quite different species. Supposedly shy and finicky plants – rare ferns, native daphnes, the only colony of wood vetch in the county of Hertfordshire – ramped along tracks we'd gouged out with a bulldozer. Violet orchids grew in thickets where it was too dark to read – then vanished when the light broke in.

Everywhere I have travelled plants have surprised me by their dogged loyalty to place, even to the point of defining the *genius loci*, and then by their capricious abandonment of home comforts to become vagrants, opportunists, libertines. I've seen ancient goblin trees develop wandering branches as promiscuous as bindweed shoots, which might equally well lope off into the countryside or jam themselves into a city wall. I've marvelled at tropical orchids living off air and mist. Plants, looked at like this, raise big questions about life's constraints and opportunities – the boundaries of the individual, the nature of ageing, the significance of scale, the purpose of beauty – that seem to illuminate the processes and paradoxes of our own lives.

And this, of course, is where the problems arise. Is it possible to think and talk sympathetically about a kingdom so different from our own without in some way appropriating and traducing it? Do we inevitably impose linguistic bondage whenever we try to celebrate vegetal freedom? Is this book's project a contradiction in terms? Our traditional cultural approach has been dominated by analogy. For at least 2,000 years we've tried to make sense of the barely animate world of plants by comparing its citizens to models of liveliness we understand – muscles, imps, electric machines and imperfect versions of ourselves. Daffodils become dancers and ancient trees old men. The folding or falling of leaves is a kind of sleep, or death. We're Shakespeare's forked radishes attempting to solve the monkey puzzle.

Metaphor and analogy are regarded as inappropriate, even disreputable, in scientific quarters. They're liable to divert attention away from the real-life processes of plants, and to end in the ultimate heresy of the pathetic fallacy, of seeing plants as the carriers or mirrors of our emotions. But I can't see how we can hope to find a place for ourselves in earth's web of life without using the allusive power of our own language to explore plants' dialects of form and pattern, and their endless chatter of scents and signals and electrochemical semaphore. In return the plant world has repaid us with a rich source of linguistic imagery. Root, branch, flowering, fruiting – we can think more clearly about our own lives because we have taken plants into the architecture of our imaginations. The trouble has been not so much with metaphor itself, as with a kind of literalism, where what is intended to be simply an insightful allusion becomes a humanoid tree or a pansy (from French *pensées*, thoughts) endowed with tender feelings. (An extreme example of this was the fashionable Victorian fad for 'the Language of Flowers', which ascribed plant species with a code of arbitrary 'meanings' which had no connection whatever with the lives of the organisms themselves.)

The great Romantic lover of plants, Samuel Coleridge, understood these tricky borderlines. 'Everything has a life of its own,' he wrote, 'and ... we are all *one life*.' He was talking about the existence of the individual inside the community of nature, but he might also have been pondering how we set our measure of the world alongside, so to speak, the plant world's measure of itself.

The diverse chapters that follow chiefly involve encounters between particular plants and particular people, and underline the point that respect for plants as autonomous beings doesn't preclude our having a relationship with them. Indeed, one alternative to viewing ourselves as natural capitalists would be to begin thinking as natural cooperators. Or as the participating audience in an immense vegetable theatre in the round. In 1640 John Parkinson, apothecary to King James I, wrote a book entitled *Theatrum Botanicum. The Theatre of Plantes*, though its subtitle – *An Universall and Compleat Herball* – gives away the staid procession of second-hand plant remedies that follows. I wanted a frame which suggested the possibilities of a more intimate, interactive relationship between our two spheres of existence; a sense of the vegetal world as protean, dissident and Learish, full of mimicry and unexpected punchlines, and a long way from abiding by anyone's stage directions. A cabaret sounded like the right kind of show.

Some of the chapters (or acts, maybe) are portraits of individual organisms – the Fortingall yew, for example, probably the oldest tree in Europe and a hapless arboreal celebrity; Newton's apple tree, whose genetic and ecological history lays waste to the gloom of Newton's physics. Other chapters are meditations on whole groups of plants – oaks, orchids, carnivorous species – whose rich cultural histories braid with their own ecological narratives. There are chapters on writers and artists – Wordsworth on daffodils, Renoir on olives, photographer Tony Evans on primulas – whose vision changed our understanding of the vitality of plants and how we might relate to it. There are accounts of some personal explorations of the Burren in Ireland and the gorges of

Crete, and what their flora says about the dynamics of vegetation, in the past and in the future. And there are introductory sections on ideas, for instance Romanticism, the role of glass in plant theatre, and the vexing question of plant intelligence.

This all sounds very serious. Plants are also fun and feisty, and I hope this book celebrates that, as well as their gift to us of different models of being alive.

HOW TO SEE A PLANT

THE SACRED LOTUS, with its bounteous white flowers and pristine leaves levitating above even polluted Asiatic rivers, is one of the most beautiful and revered plants on earth. Across 2,000 years and a swathe of cultures, it has been a symbol of purity rising out of corruption. Right up until the Maoist Revolution Chinese children were expected to memorise a lotus homily written by the eleventh-century philosopher Zhou Dunyi:

> [The lotus] emerges from muddy water but is not contaminated; it reposes modestly above the clear water; hollow inside and straight outside, its stems do not struggle or branch ... Resting there with its radiant purity, the lotus is something to be appreciated from a distance, not profaned by intimate approach.

Vegetal hygiene becomes visual beauty becomes a respectful ethical principle. Did early Chinese scientists understand how lotus leaves were able to throw off the mud they grew through, and be seen, emblematically, as a kind of moral Teflon? It's now known that the peculiar surface characteristics of the leaves make it impossible for fluids, however viscous and contaminated, to get a grip, and they simply flow away. The leaves self-clean. Today, while garlands of the starry flowers continue to be worn as Buddhist religious symbols, the engineering of the leaves has been appropriated by secular technology to manufacture a range of products patented under the trademark 'Lotus-Effect'. They include a house paint that is claimed never to need washing, except by rain, and non-stick spoons for honey. When we look at a sacred lotus, do we intuit this fundamental part of its identity, the root source of its symbolic power as well as its survival as a plant? Or just see the gentle welcome in the gorgeous petals?

While writing this chapter I persuaded the curator at Kew Gardens to let me take some lotus leaves home. They made a rather nondescript bundle, like a sheaf of rhubarb. But I had a plan for them. My partner Polly's grandkids – aged nine, seven and four – were staying for the weekend, and I wanted to know how they 'saw' the lotus, and what they made of its remarkable behaviour in contact with fluids, especially gooey ones. They enjoyed the strange velvet feel of the leaves, at least to start with. But children aren't aesthetes in that sense. They wanted action. I tried plain water first, holding a leaf horizontally so that it formed a shallow bowl, pouring in a cup of water, then gently tilting it back and forth. Globules of silver water whizzed about the leaf, occasionally coalescing, then being broken up into grapeshot by the ribs. The children burst into hysterical giggles, a sure sign of amazement. I can't believe the youngest had ever seen mercury, but he remarked, with precocious poetry, that the water roller-balls were like 'liquid metal'. Dirty water was next, and produced the same results, but nothing compared to the theatrical effect of tomato ketchup, which wriggled about the leaf like a company of disorderly scarlet slugs. Then I just tipped the whole lot off and flourished a completely unstained leaf.

Later I showed them an electron micrograph of the leaf's surface, and the rows of close-stacked, smoothly rounded pimples which are the reason even the stickiest fluids can't get a grip. But they were only marginally interested. This was not what they had seen, had *experienced*. I left them with the remaining leaves and their yelps at the behaviour of egg and golden syrup, until stimulation fatigue set in and the leaves became more interesting as sun shades and face slappers.

Seeing a plant is a matter of scale and relevance. We notice aspects which accord with our own frameworks of time and proportion, and which speak to our needs, whether aesthetic or economic. Only rarely – when we're children, for example, untroubled by such refined and narrow neediness – do we sometimes glimpse what is important for the plant itself. In this section I have looked at how the first modern humans, the Palaeolithics, saw plants; and how, working with an insightful photographer, I began myself to learn how the superficial appearance of a plant – our framed image of it – relates to its own life and goals.

1

Symbols from the Ice: Plants as Food and Forms

THE COMPELLING IMAGES OF NATURE in the caves of southern Europe are our species' earliest surviving works of imaginative representation. The galloping horses and rippling bison created by Palaeolithic artists up to 40,000 years ago are very evidently themselves, but also seem to stand for elusive abstract notions: symbolic forms, the energy of movement and creation, perhaps a world beyond the physical. What is curious, given the way that plant representations were to proliferate in future millennia, is how sparse images of the vegetable world are, and how vague. Most of the creatures painted on cave walls or engraved on bones are instantly recognisable as animals. There are a handful of images, too, that have a vaguely branching plant-like quality. But I have only seen one image that is a convincing picture of a specific, potentially identifiable flower. On a bone found in Fontarnaud Cave in the Gironde and dating from about 15,000 BC, a twig bearing four bell-like blooms rises up like a miniature maypole in front of a reindeer antler. The flowers are lantern like, pinched and cut into a V at the lip, with their stalks projecting alternately up the stem. It's a passable impression of a sprig of bilberry, or crowberry, or one of their ericaceous relatives that grew abundantly on the late ice age tundra. Foliage and fruit were food for the reindeer, which, in turn, were food for the local hunter-gatherers. If this is a deliberate juxtaposition, it's a clever and symmetrical one – prey animal and prey's forage – except for one complicating feature. When I looked at a close-up photograph of the carving, I spotted something I

Deer's head and bilberry (?)
carved on a bone from a cave in the Gironde, *c.* 15,000 BC.

hadn't seen before. Near the point at which the each bloom grades into the stalk there is a small curved line, like a breve or a closed eyelid. When I focused on it, the 'flowers' suddenly flipped, like the shapes in M. C. Escher's optical illusions. They became birds' heads and necks, or maybe a notional impression of young, suckling animals. The flower as feeder as well as food. Had the artist made a kind of visual pun, or a metaphorical image about the circularity of the food chain?

Palaeolithic artists used metaphors freely. Dark pubic triangles represented women, and probably the idea of creation. Carvers used the natural curves in cave walls to highlight the rounded bellies of animals – rock paunches hinting at fodder consumed, or calves to come – and to give them the illusion of movement in flickering light. This stands for that. The long habit of seeing resemblances and analogies has been a defining feature of our species since the dawning of the modern mind in the caves, 40,000 years ago. I can't help looking for metaphors in ice age art, any more than its creators could resist inserting them. So I may be translating this conjunction quite wrongly. Perhaps it is just the result of one artist filling in the bare space on another carver's bone. Perhaps it is another kind of metaphor altogether, an image of sleep maybe, or not a metaphor at all but a sophisticated doodle, unconnected to the world's greenery.

But Palaeolithic cave artists seemed chiefly to find vegetation visually unstimulating or short on meaning, despite the ubiquity of plants in their lives and landscapes. In the painted subterranean galleries of southern Europe you can only see plant forms by wishful thinking, conjuring up, say, a schematic tree from a few random scratch marks and ochre smudges. Cave art is overwhelmingly devoted to animals, but their habitat is invisible. Lascaux's bison walk on air. The wild Tarpan ponies in the extraordinary, 35,000-year-old 'Horse Panel' in Chauvet Cave flare nostrils, pout and whinny, but never graze. There are plenty of food animals – deer, bison, mammoths – but no food plants. There are also fish and foxes, cave bears, predatory big cats, whales and seals, a tiny grasshopper, and three splendid owls from Trois Frères (Ariège), whose expressions have the same inscrutable wisdom conventionally given to owls in modern times. The flair with which these creatures' vitality and shifting moods are captured makes it plain what caught the artists' attention, and you wonder why their own kind didn't inspire the same empathy. There are a few female figures engraved in caves and on artefacts, but they are mostly orotund symbols of pregnancy, tropes of pure fertility, and they lack the faces and individuality of the animal portraits. Plants don't seem to belong to the same cosmos.

Yet they were indispensable parts of Palaeolithic life. Hunter-gatherers were just that, gatherers as well as hunters. They foraged for berries and roots. They would have understood how the migration of their food animals was governed by the growth spurts of tundra vegetation. They worked expertly in wood, once trees had begun to return to the landscape in the wake of the retreating ice. Remains of wooden bowls, tools, shelters, even cave-painter's ladders, have been found in sites dating from the end of the Palaeolithic, circa 12,000 BC. The foragers may even have been making happenstance experiments in cultivation. Evidence of cereals, such as oats and barley, has been found in sites in Greece and Egypt. Concentrated clusters of grass pollen grains have been excavated in Lascaux, suggesting hay was gathered by the armful, perhaps for bedding. The evocative scent of drying grass may have drifted through the dreams of early artists, but not through their work.

Utility rarely seems to have been an overriding consideration in these first artists' choices of subject. Non-food animals are pictured just as frequently as prey species, and the animals featured on the cave walls are rarely those whose stripped bones, the remains of dinner, are found on the floor. As the anthropologist Claude Lévi-Strauss has famously suggested, the painters' subjects were not so much good to eat as 'good to think'. Perhaps this is one reason why plants so rarely feature. They can't be 'thought' in the same way as sentient creatures. They have no obvious *animus* or spirit. Their life cycles don't follow the comprehensible pattern of pregnancy, birth and death which all animals share. In the big bisons in Chauvet Cave – drawn with a few simple strokes that give them the bulk and energy of Picasso's bulls – you are seeing a creature whose power and thunder and fecundity have been felt in the bone by the artist. It is hard to imagine any clump of green tissue, however culturally significant, nourishing such fellow feeling.

My friend, biologist and painter Tony Hopkins, spent twenty years sketching rock art round the world. This is no longer regarded as a reliable recording technique, given the seductive opportunities it allows for fanciful interpretation. But the process of creating impressions of the images, forming reproductions of the pictures his fellow artists made tens of thousands of years ago, has given him a privileged insight into what might have been in their minds. Beyond a few comparatively recent Australian Aboriginal paintings of gourds and yams, Tony has seen no truly ancient representations of plants. His own interpretation of this, he tells me, is 'that most cultures saw plants as being part of the landscape, the same as mountains and rivers. This does not mean that they didn't think they had "spirits". But plants were not part of the palette of iconography. I think this might be because people could not see themselves in the plants, but could see themselves (or could see their shamans) transmuted into animals. Perhaps this implies that people did not think plants were "alive".' Or perhaps took them for granted, as predictable 'givens', like their habitat, or cloud formations, or their own bodies, also rarely represented in their art. Yet the absence of plants from rock art doesn't necessarily mean they were absent from the imaginative world of the Palaeolithic. They may have had meanings that were not easily accessible to representation, in the manner of scents which can be described only by similes or by reference to other scents.

Theories about the 'meaning' or, even more riskily, the 'purpose' of cave art have flourished since the first examples were uncovered in the late nineteenth century, and tend to echo the contemporary zeitgeist. The Victorians, smarting at the way this sophisticated art undermined the presumptions of civilisation, dismissed it as a collection of doodles, or the work of unconsciously gifted copyists. In the early part of the twentieth century, ethnographers with colonial models of 'primitive' cultures put the pictures down as hunting cartoons, narratives of the chase, field guides to the most desirable quarry. Or perhaps as magical aids to hunting, an envisioning and therefore 'capturing' (a word we still use about pictorial likenesses) of prey. In the psychedelic mood of the 1960s and 70s, there were theories which linked the images to altered states of consciousness and shamanic drug-driven rituals. Spears were spotted everywhere, as were genitalia (particularly dark pubic triangles, the pan-global symbol of generative power) and the pictures were interpreted in some quarters as Palaeolithic pornography, celebrations of sex and violence. In the late twentieth century French structuralists concentrated their attention on the overall arrangement of the pictures inside caves. Bison confronting horse, for example, has been interpreted as a masculine–feminine opposition, and a clue to the structure of Palaeolithic belief systems. The location of paintings in very remote parts of the caves may suggest that they were positioned close to a metaphorical portal to the animals' spirit world.

Nowadays most archaeologists are wary of grandiose, overarching theories about the meaning of cave art. Indeed, simply to ask what they 'mean', as if there were a solitary purpose and a single artistic 'language' in the Palaeolithic, smacks of a kind of patronisation, a reluctance to credit people who are manifestly already using their imaginations expressively with the same rich muddle of beliefs and feelings that have always lain behind creative art. We will never know for certain why ice age people made images, or chose the subjects they did. Jill Cook, senior curator in prehistory at the British Museum, put the dilemmas (and some of the wilder theories) of interpretation in perspective when she described an exquisite sculpture of a water bird found in a cave in southern Germany. It is only two inches long, but perfectly streamlined, as if caught in the act of diving. It may, she says, be a 'spiritual symbol connecting the upper, middle and lower worlds of the cosmos ... Alternatively it may be an image of a small meal and a bag of useful feathers.'

From the simple act of looking at the pictures, one thing is indisputable: their creators were *artists*, in exactly the sense we understand today. Their work vivaciously displays all the emotions associated with the acts of image making and image viewing: wonder, love, fear, amusement, a celebration of life, a satisfaction in the business of creation. The first reaction of most lay viewers of Palaeolithic art isn't one of anthropological interrogation but of recognition, plus a delighted astonishment that these remote ancestors saw and created in ways that are so comprehensible to us. Art was born 40,000 years ago, in John Berger's phrase 'like a foal that can walk straight away'. This isn't a matter of imposing modern sensibilities on 'primitive' intelligences. The Palaeolithic mind *was* the modern mind in embryo. That defining movement towards self-awareness, when the mind shifted, became conscious of itself and of the fact of consciousness; and an image in the memory, the mind's eye, meshed with the shadows on the rock, was also an *aesthetic* event. It meant that nature could be seen detached from itself, across space and time, and that choices could be made about how it was represented and seen.

In 2013 Jill Cook curated *Ice Age Art: Arrival of the Modern Mind* – a spellbinding exhibition in the British Museum of what is called 'portative' art, in contrast to 'parietal' or cave-wall art. The Palaeolithics scratched images on ox shoulder blades and mammoth tusks and interesting pebbles. They engraved deer on deer antlers, used the curves in animal bones to suggest perspective. Just occasionally they scraped out plant-like forms – a twiggy fork, a suspicion of a leaf. Some of these portable pieces might have been magical, carried to bring good luck on the hunt. A few have been found with the bones of their owners, as if they were tributes or grave goods. But the majority seem more light hearted, with the look of art made for whimsy or pleasure, and may have been the work of different community members, less specialist workers than the ones who fashioned the set pieces inside the caves. Palaeolithic bone whittlers made toys, trinkets, tiny vignettes. One engraved a whale with a look of sublime serenity on a fragment of whale ivory. Another, about 15,000 years ago, cut a small figure of a ptarmigan into a bleached reindeer antler, a white bird in a snowbound and plantless void. In France, at about the same time, a hunter with a sense of vision and plenty of time to spare made what was essentially the first animation projector. It is a small bone roundel laboriously shaved down to just 0.1 inches in thickness, a monumental task. On one side is an engraving of an auroch's cow, on the other her calf, whose outline beautifully catches the sagginess of young bovines. In the centre of the disc is a hole through which a thong was threaded, so that the disc can be twirled and the mother and calf made to perform a dance of transformation. It is a kind of Stone Age flick book, an ancestor of nineteenth-century moving-image machines. Most extraordinary is the oldest wind instrument in the world, a 35,000-year-old flute cut from the wing bone of a griffon vulture. Experiments with replicas show that the way the holes are positioned make them exactly reproduce the top notes in the modern diatonic scale.

Jill Cook mounted these pieces in small glass booths, so that as you gazed at them you saw the reactions of the viewers opposite you: engrossment, recognition, tears; many complex expressions of loss and rediscovery. The poet Kathleen Jamie was there, and wrote about this shared sense of time dissolving: 'Perhaps because we were Palaeolithic for such an age, the artworks we see before us are deeply, if strangely, familiar. We peer and half remember.' Kathleen and I agreed afterwards that we felt something strangely akin to homesickness in front of these extraordinary miniatures made by our ancestors, and that gathering together in a circle round these enclosed images touched some deep memory of those evenings round the cave fires.

Yet there were no plant portraits in the exhibition, despite the fact that it is on portable pieces that most of the meagre collection of early images appear. When Paul Bahn and Joyce Tyldesley reviewed all images in the European Palaeolithic era that might possibly represent or reference plants they found just sixty-eight, and of these fifty-eight were portative. Compared to the dazzling artistry of the animal portraits (probably created by an elite class of full-time painters and engravers), the exterior images are naive. Half a dozen or so are crudely naturalistic. On a baton carved from reindeer antler there are three stalks of an aquatic plant, possibly a water milfoil. A pebble from the Grottes de Cougnac near Gourdon is engraved with what is clearly a monocotyledon (the large section of plants that includes grasses and lilies). The finders suggest it might be a tulip, but the little cluster of pearl shapes above the thickly sheathed leaves look to me more like an orchid, or a lily of the valley in bud. The most gracious – and accurate – is a spray of what is almost certainly a willow, carved on a reindeer shoulder blade. The leaves are alternate, and there is a side twig that tracks the swelling at the head of the bone, a classic Palaeolithic trope. There is even a stick-beast impression from the Ariège of what is probably a deer or cow browsing low foliage, but mystifyingly interpreted by the French structural anthropologist Alexander Marshack as 'a man in the midst of stylised reeds or rushes'. And then there is the puzzle of the bilberry-reindeer collage.

Most of the rest are little more than variations on simple leaf or twig forms. Branching patterns appear often, as radiating ribs in a leaf, or a series of forks. Structuralist interpreters are reluctant to take these images at face value. The simple fork, the basic binary division, the turning of one into two, is a universal pattern in nature, as it is in the structure of thought. The bifurcating twigs may be fertility symbols, or then again feathers, or fins. All have been the subject of intense study by Marshack or his colleagues. Arl Leroi-Gourhan translates a simple engraving of a small twiggy plant with roots as a female oval symbol supporting a branching male symbol. Inverted V shapes on a carved rib from the Dordogne are interpreted as either tree buttresses or symbols of female 'entry points' (they may be both – splayed and buttressed roots are commonly seen as vaginal symbols by surviving hunter-gatherer communities in Amazonia) and the monkish hominids walking past them as men carrying sticks on their shoulders. According to who is interpreting the drawings, stick images may represent harpoons, 'male elements', or just plain sticks. Here and there on other bones and pebbles are clusters of small cruciform or star shapes. Marianne Delcourt-Vlaeminck interprets these as schematic flowers, but they may be stars, or sparks, or those optical flashes known as phosphenes that are often experienced in the dark, and especially during moments of awareness heightened by dance or drugs.

Much of the structuralist deciphering of this slender collection of plant images seems to me as fanciful as glimpsing pictures in the embers of a fire. Sixty-eight roughly scratched images, a minute fraction of what must have been created over a time span of more than 10,000 years, are not remotely sufficient evidence to tease out a lexicon of Palaeolithic foliate symbols. In the absence of any solid proof either way, I'd like to suggest a more inclusive and equally plausible account of their origins, which for me makes these early moderns feel more like the ancestors they were. The rank-and-file hunter-gatherers have a little time on their hands, maybe after a meal. They have a superfluity of stripped bones and flint burins, and are becoming familiar with the idea of making pictures from the work of the 'professional' artists inside the caves. The Palaeolithic evening class chip and scratch, some maybe working in pairs or groups, some producing little more than scribbles, like the marks young children make when they first have access to pencil and paper. The more accomplished are attempting to capture the intriguing forms that make up the green backcloth to their lives, maybe adding ciphers and clan emblems and entirely personal curlicues. The plant as a spontaneous cultural motif, rich with everyday metaphorical meaning, has its first tentative beginnings here. I'm reluctant to call it 'decorative', because of the etymologically unwarranted associations of superficiality which flutter around this idea. What I'm suggesting is that plant images and metaphors were used very *freely*, and continue to be so in the visual vernacular, as if floral and foliate growth somehow echo the dynamic processes of our imaginations. If animals have chiefly been metaphors and similes for our physical behaviour, plants – rooting, sprouting, forking, branching, twining, spiralling, leafing, flowering, bearing fruit – have, from these hesitant beginnings in the Palaeolithic salons, come to be the most natural, effortless representations of our patterns of thought.

But the physical evidence of these beginnings is slim, and if the paucity of images rather than their style is really the important part of the story, there may be an alternative interpretation. Not yet part of early culture and belief systems, kept outside the dark theatres where the human imagination took flight, undomesticated and unrevered, plants were still essentially *wild*. It was a quality and a status that was soon to be eroded by the invention of agriculture. What had been 'outside', beyond, was now 'taken in', enclosed. What had followed its own twiggy path was now drilled into our straight furrows. If you consider the whole tradition of images and conceptual framings of plants, it is striking how often they occur in spaces that have a sense of enclosure and possession.

Depiction of plants doesn't start again till 5,000 years after the end of the Palaeolithic era, and the simultaneous beginnings of agriculture in the Middle East. In Egyptian art, already well populated with birds and animals, notional plants begin to appear about 2500 BC. A thousand or so years later the pictures are beginning to be impressively accurate, and often embedded in some kind of narrative of enclosure. On one of the walls of Sennedjem's tomb in Thebes there is a painting of a fully working farm. The estate is surrounded by irrigation channels and divided up into tidy fields. In one, a man (possibly Sennedjem himself) and his wife are picking bushels of flax. In an adjoining plot he is harvesting what look like ripe barley heads with a sickle. It is a naturalistic portrait of humans in full command of the plant world but, positioned as it is in a nobleman's tomb, is probably allegorical, too – the farm already, as it would continue to be, a symbol of human life on earth, with all its stations of fruiting, harvest, death and rebirth.

At much the same time, in nearby Mesopotamia, the Genesis myth of an enclosed garden began to take shape, a model which was echoed three millennia on in the *hortus conclusus* of the Middle Ages. The first botanic gardens, created in the seventeenth century, were attempts to reconstruct the lost order of the Primal Plot: Eden itself. When they later became centres for the advancement of science and commerce, the sense that they were also botanical theatres for staging the unfolding dramas of theology and science was unmistakable. Soon botanical wonders from across the globe were being displayed on literal stages, in the bijou bell jars of Victorian drawing rooms and the great glasshouses of country houses and public gardens. Contemporary descriptions of the multitudes who came to stare at them hint that they resembled excited theatre audiences, and sometimes supplicants in a temple.

In his famous 'Allegory of the Cave' from the *Republic*, Plato tries to express the nature of the everyday experience of the physical world by imagining a group of people chained inside a cave, facing a blank wall. They watch the play of shadows projected on the wall by things passing in front of a fire located behind them, and begin to give names to the shapes and to discuss their qualities. The shadows – visual metaphors, so to speak – are as close as the prisoners get to viewing the real world.

The beginnings of horticultural order. Garden of fruit trees enclosing a flower-bordered duck pond, Tomb of Nebamun, Thebes, *c.* 1350 BC.

Plato then explains (rather self-servingly nominating the philosopher as the one character qualified to find a way out) that only by leaving the cave can people experience the true nature of reality.

Our perception and understanding of plants has, shall I say, been less black and white and more diversely democratic than Plato's allegory. But the counterpoint between 'real' plants and the shadowy forms of metaphor, and between the spontaneous, imaginative experience of vegetation and the models of scientific, commercial and priestly elites, are themes which meander through this book.

2

Bird's-Eyes: Primulas

A COUPLE OF MILLENNIA AFTER PLATO I was in a kind of cave myself –
except that, unlike a meditative Palaeolithic, I was on the inside looking
out, and at a living organism, not an image. Tony Evans had set up his
Wind Tent Mark 10 with a long view over the honey-coloured grasslands
near Shap Fell, on the eastern edge of the Lake District. The flower at
the open end of the tunnel was a bird's-eye primrose, which we hoped
to feature in a book we were making together. It was sheltered from the
wind for Tony's photo, but open to the light and the vast sweep of the
fells. Caught in this curious hinterland, in its own soil but out of its environment, it seemed sculptural, in a state of suspended animation. The
intense, focused stillness of the coral-pink flowers and powdery leaves
had the look of minerals more than plants. The picture Tony finally took
and archived is stately, a group of thin-stemmed blooms defiantly sharp
and upright against a distant haze of gathering cloud, riders of the storm
to come.

How to picture plants? Ice agers scratched stalks on bones and couldn't
resist making symbols from them. Stone Age medicine men and medieval
theologians saw metaphors of human organs in the shapes of flowers.
Is it possible to see beyond these human framings, and envisage what
William Wordsworth, writing of the bird's-eye's cousin, the common
primrose, called 'the plant itself'? D. H. Lawrence, continuing this idea,
talked of the primrose's 'own peculiar primrosy identity ... its own individuality which it opens with lovely naïveté to sky and wind and William

[i.e. Wordsworth] and yokel, bee and beetle alike'. But the quintessence of a plant can only ever be a fantastical goal, something to travel towards but never reach. Entrapped in our human brains, we can only ever see plants through our human imaginations and thought structures.

Might photography, with its receptivity to surfaces and the 'lovely naïveté' of appearance, be the least compromised route to a plant's individuality? But, then again, are the superficial details of colour and form really pointers to its character, its strategy for living, so that in them – as in the Romantics' poems on nature – 'the optical becomes the visionary'? Forgetting for a moment plants' intangible associations, many strictly vegetal qualities don't lie within what we'd normally regard as 'the optical': relationships with cryptic geological features, for instance; invisible but intense chemical communication with insects and other vegetation. Plants also have a special relationship with time that makes photography's unique ability to 'freeze the moment' of little significance. Their immobility and long periods of slow growth can make one moment in their lives appear barely distinguishable from the next. Their shapes and places of habitation, such crucial aspects of their identity, often result from historical processes involving natural and human pressures which stretch over hundreds or thousands of years. John Berger has written of the fundamental difference between photography and painting in representing time. If photography can stop time, freeze it, representational painting *contains* it, not just because its making *takes* time, and therefore incorporates it, but because it can hint at what might have gone before and might come after. Might it be possible for the photograph of a plant to have something of this painterly quality, and record not just an isolated moment but suggest the organism's past and the invisible dynamics of its life?

I'd longed to see a real bird's-eye primrose. In my late twenties a girl-friend had given me a watercolour of the flower, painted in 1778 by Louisa, Countess of Aylesford. I think it had been part of a large folder of sketches. Louisa was twenty-seven at the time, a precociously gifted artist who went on to produce twenty-seven volumes of illustrations. Her painting of the upland primrose is so ethereal and daintily done I'd assumed it was a miniature – and probably a faded one – of something altogether more substantial. The leaves are grey-green with glaucous undersides (which give the scientific name *P. farinosa*: the mealy primrose, the floury flower) and the tiny five-petalled blooms sport in their middle a bright circle of yellow – the bird's eye. Louisa had painted her specimen in the rock garden of her family seat, Packington Hall in Warwickshire. I wanted to see it in its wild motherlode, up in the pastel landscapes of the northern limestone, its sole location in Britain. Geoffrey Grigson wrote of how 'the small neat flowers decorate every bank, every slope, every corner between the grey lumps and outcrops of limestone. Finding them for the first time, a southerner feels like a plant collector on the Chinese mountains.' I'd felt more like an ignorant grockle when I first searched for them, arriving in the Dales weeks after they'd finished flowering.

So the flowers I saw on Shap Fell with Tony were my first. And they were to the life, in size and presence, the species in the Countess's picture. They seemed to have been formed out of the very stuff of the rock, the calcareous farina of the limestone, the pink fragments of metamorphic rocks. I was so seduced by this conceit of stone suffusing tissue (thinking perhaps of the way a vineyard's *terroir* flavours its grapes) that I never even noticed the nominative bird's eye, the same colour as a forget-me-not's, which sits at the centre of the flower and is the beacon that helps perpetuate its kind. I didn't understand its role in pollination then but it sits in plain sight in Tony's picture, a reminder that time changes understanding too.

I'd met Tony Evans four years earlier, when we'd been commissioned by *NOVA* magazine to do a feature on Britain's vanishing wild flowers. Tony would take the pictures, and I would write the words, though the editor seemed uncertain about who should be the lead partner and make the choices of species. It didn't matter in the end. We got on from the start, and the flowers we decided to highlight emerged by a kind of symbiosis, Tony sponsoring the good lookers, me the ecologically meaningful. It

Depths of focus: spring flowers in a Dorset hedgebank, Tony Evans *c.* 1974.

wasn't long before I'd persuaded him to collaborate on a book I'd been musing about for years, on the cultural history of the British flora. It was the beginning of a working partnership as comfortable (and just occasionally as prickly) as that between a butterfly and a bramble, and a friendship that lasted until Tony's premature death in 1992.

At the time we met Tony was one of Britain's most successful commercial photographers, sought out by advertisers and magazines for his obstinate perfectionism and wildly idiosyncratic solutions to visual challenges. Some of his most inventive pictures have become part of the national folk memory. His cover picture for the *Radio Times* for the 1976 Royal Variety Performance was of a corgi emerging from a top hat, which involved days building a hidden frame in which the dog could comfortably endure Tony's hours of meticulous preparation. He took the definitive portraits of Ray Charles and the fidgety Alfred Hitchcock. He spent weeks growing onions and tomatoes compressed in glass cubes so that he could cut them in square slices for a processed cheese advert. His record work time for a single photograph, I think, was the eleven days he took waiting by a cockle bed in Lancashire for the exact combination of tide and light needed to tempt the molluscs up into visibility. His expenses claims were legendary. But he was getting tired of the contrivance and pressure involved in commercial work, and feeling a desire to work more naturally – in fact, *on* nature. I sensed that his patience and Martian eye for the visually strange fitted him up very well for venturing into the world of plants, and for us working together during long summers on the road. We shared a similar sense of humour and a taste for gadgets. His working Dormobile contained, in addition to a dozen aluminium boxes of photographic equipment, an entire set of Ordnance Survey maps, an altimeter, various sunshades in different tones, a set of ski-stick ends for anchoring tripods in bogs, and a fridge which always carried a stock of decent white wine. Lunches were always a congenial prospect. But our inputs to the project were going to be different, beyond the simple provision of, respectively, pictures and words. Tony would inject the intense attention to the living plant that I so often bypassed in my concentration on its associations and history. For my part, I hoped I might stretch Tony's already encompassing vision beyond its focus on form and composition into imagining them in their native habitats. In reality we learned from each other, and over six summers unexpected new perspectives began to emerge – especially the possibility of expressing the past and the abstractions of metaphor in that epitome of the literal and the present moment, the photograph. 'You can't photograph what isn't there, Rich,' was Tony's exasperated response to my more fanciful suggestions. But he could, and, plant by plant, he did.

So for a few weeks every spring and summer between 1972 and 1978 we went on the road, following the primrose path across Britain. There were a few shaky moments for me at the beginning, when I felt that Tony's technical brilliance in catching the sheer surface dazzle of flowers might preclude any more nuanced insights. But I think that was just a touch of professional envy. It soon became a journey of discovery for both of us, into corners of Britain and perspectives on growing things we had never experienced before. We travelled, in no particular order, from the Sussex Downs, where we buried ourselves in the intricate warp and weft of downland flora while listening to Virginia Wade tease her way to victory in the 1977 Wimbledon singles final, to the far north-west of Scotland, where Tony photographed yellow saxifrage and rose-root under his pink Mark 2 parasol. In the limestone country of the Burren in County Clare he took one of his most fully accomplished pictures, a confetti of burnet rose petals floating in what resembled a limestone font, surrounded by a chaplet of rose bushes. A landscape condensed into a single boulder.

Tony was stretching the boundaries of plant photography in studies like this, with their melding of pictorial and ecological composition. Before he began his work in the 1970s, portraying a plant without artificial lighting was unusual, and combining foreground floral detail with habitat context barely heard of. Quite unprompted, Tony started to photograph ecologically, using, for instance, the layers of light and shade in dense vegetation to provide a kind of natural highlighting. His close-up of a square foot of Dorset hedgebank uses deep focus to pick out dog's violets in the shadows and primroses and celandines in the community's top layer. It is composed like a painting by Dürer. And looking at it afresh today, I spotted two small wild garlic leaves hidden beneath the celandine that I'd never noticed before ...

In June, in the limestone grassland of Lathkill Dale in Derbyshire, we found a large colony of wild columbines, white and blue. They were an unsought bonus, and Tony set to work on them under the blazing sun. He had me scissor his jeans into shorts while he worked, and stayed in the same spot for the next six hours. I watched from the shade with increasing anxiety as, bare backed, he rotated round the columbines like a sundial's shadow, or an attentive insect.

On the road we developed a slightly playful sense of plants as currency: I paid my dues to the book in stalks and clumps, Tony in the delicacy of petals. We also began to experience weather from the plant's point of view. Tony got regular pinpoint forecasts from the nearest airbases to pick the most promising times for pictures, but we saw weather's effects translated through responsive plant tissue every day. This was the impetus behind the development of the wind tents. Tony needed an enclosure for sheltering plants in gusty weather which didn't separate them too much from their habitat. An open-ended, low-lying bivouac seemed the answer, but he went through many prototypes – large mouthed, small mouthed, wooden and aluminium framed – before arriving at the version in which he photographed the bird's-eyes.

Tony had Zen levels of patience and attention. He was never a high-speed snapper, taking multiple images and postponing aesthetic judgement till he'd seen the processed film. He had built the final image in his mind before he took a single shot. On these vigils we sometimes had glimpses of the methodical way that flowers opened. How complex, perhaps mathematically pre-set forms, unwrapping themselves over the course of hours or days, invariably appeared as a recognisable type but in individual blooms, as subtle variations on that model, customised by the fuzzy logic of growth.

Tony's perfectly achieved representation of this process, an incomparable essay on plant form and its inexplicable idiosyncrasies, happened fifteen years after we finished our book, and just a few before his death from cancer. He photographed, in close-up and high resolution, the stages through which a single bloom of Florentine iris passes during its opening. He sat with the pregnant bud for twenty-four hours, in a studio rigged with alarm clocks in case he nodded off, and took six studies. The head of the bloom opens symmetrically at first, like the mouth of a fish sucking at the surface of the air. Then what will become, in the final photograph, the right hand 'fall' (the term for the skirt-like lower petals – there are three of them) tilts down by itself, still scrolled up, followed a few hours later by another on the left of the picture. Only when all three are fully lowered does the entire flower – crown and falls together – unfurl into its glorious, baroque orb of pale violet, thinly streaked with reticulations of a darker lavender. I have the set of six prints framed on my wall, a reminder of Tony's vigilance, and a soliloquy on the quiet unexpectedness of floral rhythms.

Back on the road the daily round involved these essentials. Moments of soft light: nothing made soft petals crinkle or leaves take on the shine of overheated skin like fierce sun. Long breaks for a formal picnic spread out on a tablecloth, wherever possible next to a stream, so that a bottle of Sauvignon could be cooling in the shallows, and we could review the morning's work. It occurs to me now that, with our enormous backpacks of camera gear and creature comforts, we were like nothing so much as two Victorian plant hunters, intrusive strangers dropping into other communities' floral landscapes, passing lofty judgements on their habits and habitats, and then wafting on with our literary and visual trophies. At Oxford we fraternised with native and colonist alike. In Magdalen Meadow Tony got in amongst the swaying tides of snake's-head fritillaries to frame the sultry, serpent's-skin flowers against the honeyed stone of the college tower. Later that same day, we logged Oxford ragwort growing in a scree of broken lager bottles outside the public conveniences in St Clements. *Senecio squalidus* had been brought to Oxford's Botanic Gardens from the slopes of Mount Etna in the 1790s, a trophy of some don's Grand Tour. But it had escaped, and slummed its way through the city, relishing Oxford's venerable stones' likeness to volcanic rubble.

In our first spring Tony came back from Scotland (we didn't take all our research trips together) with an extraordinary image of marsh marigolds. We'd been talking about how to illustrate the flora of what is known as the Atlantic period that set in about 6000 BC, and during which an increase in rainfall and a rise in sea level submerged parts of the great summer forest that had succeeded the glaciers. Plants of swamp and bog flourished. Tony shook his head at the impossibility of capturing any living representation of this vanished sogginess. Then he went to the moors of Inverness-shire and brought home a photograph of an ancient pine stump – it might almost have been fossil pine from the Atlantic era – heraldically studded with five marsh marigold flowers, kingcups rampant. They were rooted in damp crevices and pits in the rotting wood, portals into the boggy northern past, and Tony had achieved what he always insisted photography was incapable of, showing something that was not there.

For me it was another primrose, the oxlip, that brought our meandering search for the full picture most sharply into focus. The true oxlip, *Primula elatior*, had pushed its way into my life in unexpected ways ever since I became seriously interested in plants. It's the most gracious of all the European Primula species, bearing clusters of primrose-yellow cups hanging to one side of the stem, in the same style as bluebell flowers. And for those with a nose for it, the blooms give off a fugitive scent of apricot. In Britain at least, it also has a quality which I've always found irresistibly romantic, of being fiercely loyal to place. It only grew, or so the books I read in the 1970s insisted, in a cluster of ancient woods in a roughly oval-shaped area of East Anglia covering west Suffolk, north Essex and south Cambridgeshire. Its cloistered distribution came to be an existential definition of the territory.

The concept of ancient woodland was a radical one in the 1970s. The popular wisdom was that woods were human artefacts, and could only begin their lives by being deliberately planted. The idea that they continually regenerated themselves, and in some places had survived on the same site for thousands of years, was heretical, and is still uncomfortable to those brought up in a culture of human dominance. One consequence of this continuity is that there are suites of flowering plants, finicky in their habitat requirements and not good at colonising new sites, which grow only in such places. It's not clear why this should be so. Most of these so-called 'ancient woodland indicators' are perennials, and don't reproduce well by seed. They may also have evolved associations with the micro-organisms in old-forest soils, or with the mycorrhizal fungi symbiotically entwined with the trees' roots.

The oxlip is one such indicator species in East Anglia. In the 1970s I used to go on oxlip hunts, seeing how often I could find the plant in woods which looked, from their names and quirky outlines on the map, as if they might have ancient provenance. My foragings weren't much more than ego-massaging treasure hunts, but they paid off in ways I couldn't have expected. Oxlips proved to be doughty communards. They stood their ground, often against the odds, and sometimes even when the wood in which they'd originally grown had vanished. I found colonies in full flower in Easter Wood in Suffolk, just days after Easter Sunday, even though they were sharing it with grazing cattle. They spilled out over the roadside verge alongside a thicket called Haws Wood, the rest of which had vanished under a pall of conifers. I found one colony mysteriously tucked under a hedgebank and alongside a track leading to a wood. Later, on a map from 1783, I saw that the track once lay inside a part of the wood which had been grubbed out in about 1800, leaving the hedgebank as a relic, a woodland ghost.

Then one spring I was walking through a small hornbeam wood I'd known since I was a child near my home in the Chilterns, and spotted a couple of oxlips in flower by the path. I think I was confused more than astonished. So cut and dried was the oxlip's supposed confinement to the woods of East Anglia, fifty miles to the east, I assumed I must have made some kind of slipshod mistake. I'd walked through this wood every April for thirty years and couldn't have missed something so conspicuous. So they must have been obscured by now-vanished foliage, or sneaked in behind my back, or I'd simply misidentified them. The last possibility was easily ruled out, and everyone I took to see the plants confirmed they were true oxlips. But talking my find over with some neighbours I heard a rumour that a retired vicar who lived in the next road was fond of planting out wild species in provocative places. I liked the idea of the Reverend Moule buzzing round the countryside on his antique autocycle with its panniers stuffed with sheaves of feral vegetation, like some botanical evangelist.

The flowers appeared again next spring, and I began to wonder if they were the only two local specimens. On the slightest of hunches, I wriggled under the barbed wire of the biggest adjacent wood. I could scarcely believe my eyes, or my luck. There were oxlips blooming amongst the budding shoots of woodruff and yellow archangel all over the south-facing slope. I eventually found them in four pockets of old woodland in an area of about a square mile, more than 200 plants in total. When I looked at the First Edition Ordnance Survey map (circa 1820) of the area, I saw that the four woods were all featured, and were part of a cluster of eleven whose outline and close jigsaw formation suggested they had been carved out of what was once a larger tract of continuous forest. The local oxlips began to have the look of an oasis population, an isolated relic perhaps of a time when the woodlands of East Anglia and the Chilterns were joined, 6,000 years ago. The thought of a tract of primordial forest joining the two places I knew best in the world made me feel strange, and transported.

Botanical notables came to visit the plants, pronounced them authentic and almost certainly native. Two dots confirming their presence appeared in the new edition of the official *Atlas of the British and Irish Flora* (2001), my one and only original contribution to botanical science. But what affected me more at the time was the way their presence had subtly changed my sense of where I lived. Right through my late childhood and teenage years these woods had been arenas of the present moment, stages for the quotidian and unreflective sensory excitements of adolescence. I stalked them ritually, touched their trees, left buried messages for teenage crushes, scoured them for fragments of reputedly crashed World War II planes. Now they and the place I called 'home' had a deep history, mapped in the lineage of these apricot-scented bellflowers. It was seized by the irrationally Romantic feeling that the oxlips were using me to retrace their ancient connections.

*

Tony and I needed more concentrated clusters than these Chiltern tufts, so we went looking in the Bradfield Woods in the heart of Suffolk's oxlip country, one of the plant's classic sites. The auguries weren't good. Many of the flowers were over, or had been chomped off by pheasants, and I was embarrassed that I hadn't done my preparatory work. When we did eventually find a few clumps large and fresh enough to satisfy Tony's demanding pictorial criteria, I was tempted to disappear, in some relief, on a private hunt for some of Bradfield's other specialities. (It has a genuine wild pear tree.) But Tony needed to know *why* I wanted to feature oxlips, what they meant to me and to the story we were telling. So I tried to explain my rather clingy attachment to them, and their clingy attachment to ancient sites, and the way they responded to the increase in light after coppicing, and how their companion plants did too, and how it had not been until the 1840s that they had been recognised as a separate species ... And as we talk, Tony sizes up the flowers around us. He chooses a luxuriant clump, with the flowers in a pyramidal, almost tent-like canopy above the leaves. He anchors the tripod, like a stout umbellifer, and takes the camera down to woodland floor level. After a while he finds in the viewfinder the narrative picture he will take: a dozen or so flower spikes, deeply grounded but rising like a sheaf in the left foreground, tangled up with dead twigs, young meadowsweet leaves, and a protective tussock of grass. In the background, caught by the wide-angle lens, is a fragment of cut-over coppice, and beyond that the new poles, still leafless, shearing towards the clear spring sky.

Without really planning it, our meandering progress across Britain moved in a vaguely northern direction, homing in on the poetical landscapes of the Lake District, where Coleridge and the Wordsworths had meditated on primroses and daffodils. We eventually hunkered down in a hotel not far from Grasmere. In Little Langdale nearby, Tony spent a whole day up to his waist in a mere to photograph white water lilies at the eye level of an approaching dragonfly. In the picture the foreground lily flower is so close to the camera that you can see the pollen on its stamens. But dozens more are visible with fly's-eye sharpness deep into the photograph, a vision of insect opportunity. Another evening we made our way up a hill nearby and found what I can only describe as a sundew meadow. Where the hill flattened out into a patch of bog, about a hundred square yards of these insectivorous plants were swarming amongst barrows of red-tinged sphagnum moss, and the whole ensemble was glowing in the setting sun. The beads of sticky fluid that tip the sundew's spines to entrap visiting insects seemed to be acting like prisms, and the bog was sparkling with minute and ephemeral rainbows.

Tony's pictures exist in two planes of time and understanding for me. I can remember the circumstances in which they were taken, and what I felt at that moment. I also have the finished pictures, forty years on, and the knowledge I have now gives them depths of meaning that I never glimpsed at the time. Lying on our fronts on the moor and seeing the sunset through the orbs of dew was an extraordinary visual experience, but not much more. Now I see Tony's picture of the plants' leaves, red and glistening on the hillocks of sphagnum, as a portrait of a softly glowing vegetable foundry, where the sundew leaves forged sunshine and insect flesh into another kind of energy.

There is no record of the Wordsworths (William the expert in the refractions of memory, Dorothy, I sense, the potential photographer) ever seeing a bird's-eye primrose, though they grew on limestone outcrops near Grasmere and had a local Cumbrian name, 'bird een'. But on one of their strolls with Coleridge they found a tuft of common primrose in a favourite spot that was to become significant in their lives. On 24 April 1802, nine days after their encounter with the 'dancing' daffodils that led to William's famous poem, Dorothy notes in her diary: 'A very wet day. William called me out to see a waterfall behind a barberry tree. We walked in the evening to Rydale, Coleridge and I lingered behind. C. stopped up the little runner by the roadside to make a lake. We all stood to look at Glow-worm Rock – a primrose that grew there, and just looked out on the road from its own sheltered bower.' William had always been drawn to plants in this kind of situation, and their demonstration of grace under fire. The Glow-worm Rock primrose was, in Molly

Mahood's words, able 'to make a home for itself in the bleakest spot, its roots anchored in a crevice's minute parcel of soil, its leaves spread low in a wind-resistant rosette, its light, bright, flowers free to move on their flexible stalks – an embodiment, all in all, of the Wordsworthian virtues of delicacy, independence and fortitude'.

Over the years that followed William returned often to this plant, in life and in his writings, and its survival seemed a kind of domestic anchor for him. In 1808, with his brother John dead and Coleridge gone from the Lakes, he wrote the poignant 'Tuft of Primroses':

> What hideous warfare has been waged,
> What kingdoms overthrown,
> Since first I spied that Primrose-tuft
> And marked it for my own;
> A lasting link in Nature's chain
> From highest heaven let down!

The final version, entitled 'The Primrose of the Rocks' was composed twenty years later, by which time the primrose's associations, reflecting William's own strengthening Anglicanism, have become as much theological as ecological, and the poem declines into pious Sunday School parable.

> The flowers, still faithful to the stems,
> Their fellowship renew;
> The stems are faithful to the root,
> That worketh out of view;
> And to the rock the root adheres
> In every fibre true.

It was left to Dorothy, still of sharp and allusive eye, to nail the primrose decisively back onto the rock and the hard realities of natural and familial survival. The title it bears in a copy she had made herself is: *Written in March 1829 on seeing a primrose-tuft of flowers flourishing in a chink of rock in which the primrose-tuft had been seen by us to flourish for twenty-nine seasons.*

In the fells north of Grasmere it had taken Tony and me ages to find the bird's-eyes. Geoffrey Grigson's description of their decorating 'every bank, every slope, every corner' was either wildly exaggerated or out of date. Having never seen the plant in the wild, I had no idea of where or what to look for, expecting them, I think, to be sprouting like tufted alpines on the bare rock. We traipsed over sheep pastures and rocky outcrops, and scanned road verges from the van. Eventually we found them by chance, growing in lime-rich trickles through the peat – a distinctive niche I soon learned to recognise. So up went Tony's meticulously refined wind tunnel in the bleached grass, and the picture was taken. Looking at it now, and putting aside my fanciful idea of the bird's-eye as a distillation of limestone, I find the pastel wiriness of its stems, held tight against the vaporous background of the hills, compelling and strange, as if the protection Tony had given them from the Pennine winds had, paradoxically, revealed their animation by it. And I can make out the bird's eye that I missed when I was only feet away. It is a true picture of the 'primrose itself' (lacking only perhaps the squidginess of the ground) that transcends that ephemeral moment when the shutter clicked. It seems a plant as defiantly and immemorially in its place as the 'Primrose of the Rocks'. But as a photograph it also records that specific moment and unlocks its mordanted memories; and every time I look at it I remember the migrating lapwings that were passing overhead, and the grass scented as if it had just been mown, and Tony, watching intensely from inside his cave.

WOODEN MANIKINS:
THE CULTS OF TREES

The Palaeolithics' dawning interest in the metaphor of the branch was picked up later in the Middle East, and it's not without irony that the first image of a significant tree is from Mesopotamia, springboard of the agricultural revolution that was to destroy most of the world's forest over the next 5,000 years. It's on a Sumerian seal from circa 4500 BC and shows a tree of life between two gods. The female goddess is probably Isis, and she has a snake, a symbol of water, beside her. The tree is smaller than the gods and appears to be on some kind of stand, but it is obviously a palm, with symmetrically arranged leaves and two prominent drooping dates. Whether the palm was regarded as a minor god itself, or just one of the gods' sacred accoutrements is uncertain. Almost identical arrangements of layered branches, occasionally with date-like appendages, are carved round the fonts in many Christian churches, again adjacent to sacred water, and are usually regarded as representations of Eden's Tree of Life.

It was just such an idealised notion of 'the tree' – a monumental, branching manikin (children draw humans as simple trees, just as they paint faces on flowers) – that placed this group of long-lived plants at the centre of many creation myths and models of the universe, especially in societies which had already generated their own branching and layered hierarchies.

Joining earth and sky, capable of outliving not just individual humans but whole civilisations, they are obvious symbols of the more-than-human, and many religious narratives have played with analogies between the arboreal seasons and spiritual cycle of death and rebirth. China has the Kien-mou, the 100,000-cubit tree of life. The Buddhist Tree of Wisdom's four boughs are the source of the four rivers of life. One of the best known is the Norse 'World Tree', Yggdrasil, usually imagined as an ash, and first appearing in Norse legend in the early medieval period, though presumably of older origins. Yggdrasil is one of the more intriguing trees of life, since it has some relation to the real ecology of the forest. In the myth it is the focal point of the transmutation of the sun's energy and of a set of reciprocal relations with other organisms.

As for the Tree of Life in the Old Testament, it has no such creative role, and no real-world botanical model. It certainly wasn't a bearer of apples, a species unsuited to the parched environments of the Holy Land. But a sense of its sheer *treeness* – its durability, rootedness, continuity, powers of regeneration – is emblematic of the symbolic power of all woody growth. And as a motif, it crops up time and again in Christian mythology. Its complete adventures, in a myriad local versions, has the tree supplying most of the wooden artefacts of Christian iconography. The story starts when Seth, Adam and Eve's third son, returns to the Garden of Eden – still there apparently, like the abandoned estate of a derelict house – and begs some seed of the Tree from the angel on duty at the gate. He plants it in his father's mouth as Adam is dying. It grows from his mouldering body into a tree which metamorphoses across species and across sacred history. It supplies planks of 'gopher wood' (possibly cypress) for Noah's ark, and an unspecified branch for Moses' omnipotent staff (which at one point changes back into a serpent). Later it is a cedar felled for use in the construction of King Solomon's temple, and a wooden bridge for the visiting Queen of Sheba to cross its encircling moat. When the temple is destroyed, its timber, after more fantastical coincidences and transformations (it occasionally reappears as a plank in Joseph's workshop), ends up forming the beams from which the Cross is made. The tree of life becomes the tree of death, and then the wooden symbol of redemption.

An elaborate portrayal of Yggdrasil, the Norse Tree of Life, from a seventeenth-century Edda manuscript.

It is also the form of tree growth – the repeated forking and dividing, the way in which the structures of bodily trunk, branch and leaf echo each other – that has helped make trees such universal symbols. Branching is also the pattern of the lineages of living things. The biblical Stem of Jesse is a tree, a family tree. Each 'begat' is a node where a new branch or twig emerges. Two thousand years later Charles Darwin, whose ideas challenged the very core of fundamentalist Christian theology, schematised the progress of evolution in a roughly sketched family tree in his notebook for July 1837. It bears a remarkable resemblance to the twiggy scrawls on Palaeolithic bones. Later he included a more refined sketch as the one illustration in *The Origin of Species*. Kingdoms, individual dynasties and families, sprout out as branches, and occasionally die out or divide again into newly evolved groupings. Darwin himself wrote an eloquent commentary on his tree, abbreviated here:

> The affinities of all beings of the same class have sometimes been represented by a great tree. I believe this simile largely speaks the truth. The green and budding twigs may represent existing species: and those produced during former years may represent the long succession of extinct species ... As buds give rise by growth to fresh buds, and these, if vigorous, branch out and overtop on all sides many a feebler branch, so by generation I believe it has been with the great Tree of Life, which fills with its dead and broken branches the crust of the earth, and covers the surface with its ever-branching and beautiful ramifications.

Darwin's Tree of Life is both a literal description of the development of a real tree through time, and the grandest of metaphors about the whole of creation.

Our modern preoccupation with the tall and stately and decoratively antique was a later development, coinciding with the period of history when trees began to be owned, and used as symbols of status and wealth. At the same time they began to be judged by human values. The deliberate creation of plantations replaced natural regeneration as

Charles Darwin's notebook sketch of the evolutionary Tree of Life, 1837.

a means of producing new trees. The straight column – architecturally classical and commercially valuable – came to be admired in preference to the gnarled and self-sprung native. The fact that trees have perfectly adequate reproductive systems began to pass out of popular conscious-ness. We now believe we have to plant them to guarantee their presence on the earth. A thoroughgoing anthropomorphism continues to infect our judgement of their form, too. Natural signs of ageing are interpreted as disease, and disorderly growth as an indication of inferiority. Both are frequent preludes to arboreal cleansing.

None of this bore any relation to how I first experienced trees as a child, hunkering down in the root pits of wind-thrown chestnuts and spending days up in the tangled attics of cedars. I saw their living interiors, not their painterly surfaces. In the chapters that follow, using some half a dozen tree species, I have set down images and cultural expectations against their own ingenious, resilient and often rowdy behaviour.

3

The Cult of Celebrity: The Fortingall Yew

IN THE EARLY 1970S, an ex-soldier, wanderer and freelance conservationist called Allen Meredith began having mystical dreams. A group of people in long gowns and hoods were sitting in a circle, instructing him to look for the 'Tree of the Cross'. 'It was called by another name,' he remembers, 'but I knew it was a yew.' There were doubtless many dreamers of Druidical figures and sacred trees in those psychedelic years, but Meredith acted on his visions. He became obsessed with the yew, began to believe that it held crucial lessons for humankind and, finally, had a revelation that it was the true Tree of Life. Almost single handedly he revived a fascination with the yew – and especially its longevity – that had begun three centuries before. His arguments, together with the evidence he began collecting, persuaded many tree scientists that ancient churchyard yews were older than all previous estimates by some thousands of years; and that one in particular, the Great Yew in the churchyard at Fortingall in Scotland, might be the oldest living thing on earth.

Spending the first half of my life in southern England's chalk country, one of their favourite natural haunts, I grew up with yews. Their seedlings, planted by thrushes, bristled impertinently on the hallowed downland turf. Bigger trees hunched amongst the grey-trunked beeches like dark Jack-in-the-Greens. Sometimes they grew to maturity, but they never looked old.

Churchyard yews have always seemed different. They have an aura of antiquity, distilled from sombre foliage and dark wood; a sense of being relics or echoes of some forgotten business, even when they are obviously not old at all. There was a yew in the parish churchyard in my home town of Berkhamsted. At 300 years old it was a mere stripling, and our teachers told us that the mound on which it was planted may have held the corpses of the town's plague victims. They never said why. We'd dash over the miniature barrow for dares, tempting up ghosts and adult disapproval. I had no idea then that yews were once held to have vampiric relationships with the dead, or I would have kept my distance. 'If the Yew be set in a place subject to poisonous vapours,' wrote the fanciful botanist Robert Turner in 1664, 'the very branches will draw and imbibe them, hence it is conceived that the judicious in former times planted it in churchyards on the west side, because those places, being fuller of putrefaction and gross oleaginous vapours exhaled out of the graves by the setting sun, and sometimes drawn by those meteors called *ignes fatui*, divers have been frightened, supposing some dead bodies to walk.' (A curious inversion of this belief survived into the twentieth century. A German medical professor at the University of Greiz, Dr Kukowka, claimed that yews emitted gaseous toxins on hot days which could bring on hallucinations in people sitting under their shade.) But our yew didn't seem a mystical or discomfiting presence, just a rather congenial civic ornament. Stuck in the centre of the town, close to a string of pubs, it was a natural gathering place for revellers on New Year's Eve, and at midnight during the storm of 31 December 1976 I watched the thin, fluent twigs blowing like black bunting over the crowds in the High Street. Whatever else it might have been it seemed a companionable sort of tree.

Yew is an ancient citizen of the temperate zone, with seven species spread between Asia and Central America. The European species, *Taxus baccata*, has grown from the coast of central Norway to the mountains of north Africa for a couple of million years, give or take the odd ice age. In Britain, archaeological remains have shown that yews were widespread between and after the glaciations. A 250,000-year-old yew spear found at Clacton in Essex is the world's oldest surviving

wooden artefact. In later periods climate change restricted its distribution. Immense trunks and stumps have been discovered buried in the
Fenland peat, drowned by the rising sea levels about 6000 BC. Those
on dry land began to vanish as soon as the first farmers arrived 1,000
years later, hoicked out because their toxic foliage poisoned cattle. The
syllable 'yew' echoes in several European languages – *iw* and *yw* in
Welsh, *uwe* in Dutch and *if* in French and German – suggesting the
tree retained some kind of continuity of meaning as settlers moved
across the continent. But the persistent and popular idea that yews were
'sacred', the revered objects of prehistoric cults, has little hard evidence
to support it. There are a handful of Bronze Age rock carvings of what
might possibly be coniferous foliage, and some unreliable anecdotes by
Roman commentators, but that is about it. Yew does not even figure as
a named tree (and therefore an important geographical feature) in a
survey of Anglo-Saxon boundary descriptions for the south of England.
Yet it was plainly a tree of some significance, as a landmark or meeting
place, because hundreds of prehistoric yews are still alive, presumably
in their original sites. A few may have survived because of the inacces-
sibility of their habitat, like the undatable dwarfs clinging to the vertical
cliffs of Yew Cougar Scar in the Yorkshire Dales. But most are mysteriously situated in the yards of churches thousands of years their junior,
which is at least a circumstantial hint of some past sacred role. These
survivors, now their ages are beginning to be understood, represent a
serious challenge to many of the assumptions that have traditionally
been made about the possible ages of trees. The oldest churchyard yews
may well have sprung to life not long after the beginnings of settled
civilisation in Britain, and show no signs of approaching demise. But
there they continue to sit, topographically entangled with the much
younger buildings of our official religion; the wild and serpentine next to
the stolid and devout. Whatever ancient dance is going on here between
biology and the social order, the yews themselves consistently refuse to
behave like solemn monuments, and the existence of these idiosyncratic
trees on holy ground poses questions beyond the likely date of their
origins, especially why and how did they come to be there.

The first veteran I saw for myself was at St Mary's Church in Selborne, Hampshire, when I was working on a biography of the village's most famous son, the naturalist Gilbert White. The Selborne yew was a local landmark, but certainly not humbling or cathedral-like or conforming to any of the other clichés with which the individuality of old trees is so often smothered. It was not even particularly big, except for its massive girth – some twenty-eight feet when it was first measured – and grew modestly on the south-west side of the church. When Hieronymus Grimm made an engraving of it for the first edition of White's *The Natural History of Selborne* in the 1780s, he pictured a distinctly stumpy growth, pollarded right back to the height of the surrounding cottages. Two centuries on it was squat and fulsome, with the repose of a country alderman. What I loved most was the interior of the trunk. Old yews are almost invariably hollow, and the inside surfaces of the Selborne tree were patched with a satiny sheen of lilac and grey, like the lustre of mother-of-pearl. It was a cheerful tree. It even had a seat round it, from which to watch the world pass by.

But the yew which trumps them all, perhaps the most celebrated and provocative tree in all Europe, is the Great Yew at Fortingall in Perthshire, located by the Victorian church in this tiny village. The faithful – and it isn't extravagant to say such trees have disciples – believe it has been growing here for at least 5,000 years. If so, it was flourishing before the making of Stonehenge and the digging of the Neolithic burial chamber at Maeshowe in the Orkneys. The Great Yew is the supreme example in Britain of the mysterious conjunction between ancient tree and sacred site, and the implications haven't been lost on modern pagans. The Christian church, they suggest, appropriated sites of earlier tree cults, and the presence of an old yew by a church is a sure sign of the Old Religion. In consequence a florid New Age folklore has begun to blossom around ancient yews. The Fortingall tree in particular has become a kind of tree goddess, revered by Druid revivalists, nostalgic wood folk and patriotic Celts. Even some free-thinking Christians have begun to mythologise it. Jesus, the legend goes, visited it during his 'lost years'. A starburst of ley lines – from the Holy Isle of Iona to Montrose (Mount of the Rose), from Tobermory's Well of Mary to Marywell on the

The Great Yew, Fortingall, in Perthshire. The circumference of the tree in the eighteenth century is marked out by posts.

coast, from Eilean Isa (island of Jesus) and the holy island of Lindisfarne – converge at Fortingall, the *axis mundi* of alternative Scotland.

Now churchyard yews everywhere are being anatomised, measured, mapped, blessed and danced round. Some acolytes see the species as the living type of Yggdrasil, the Norse 'World Tree'. A few have glimpsed in the Indo-European root of its name – 'iw' – a more than coincidental echo of 'Iawe', the Hebrew name for Jehovah, and have christened the yew 'The Lord's Own Tree'. It's a mighty weight of symbolism for a short and rather common tree to bear.

I went to see the Fortingall prodigy for myself one mid March, when the surrounding countryside was still in brown tweed. I knew old yews well enough not to expect a skyscraper, but I was looking forward to the tree frothing over the road, its spring shoots the first new green of the year. What I wasn't prepared for was the diminutive tuft, no taller than a hedgerow hawthorn, tucked under the lee of the tiny church. Nor for the fact that the Great Yew is in a cage. This is to keep us, the itchy-fingered public, out, not the tree in; at least that's the story on the noticeboard. When the yew was first 'discovered' in the mid eighteenth century, it had a bad time from souvenir hunters, who hacked pieces from the already collapsing trunk until it had effectively turned into two separate trees. By the end of the century, the gap between them was wide enough to carry a coffin through.

Squinting through the bars and reading the captions has become, alas, the 'yew experience'. You feel voyeuristic, as if you're peering through a door hole in Bedlam at an inmate slumped in the corner. The yew seems hunched as much by the enclosure as by its own ageing timber frame. The northern half – a sheaf of thick knotted stems, each as thick as a sheep – has a few thinner branches which lope across the pen, and then stop dead at the fence. The southern trunks are propped up by crutches, and here and there by the wall itself. The interior is too dark to make out its texture, and the trunks seem to be regressing to the quality of rock, not wood. But I could see the circle of posts that had been hammered into the ground to trace out its earlier, more extensive circumference. Twenty people could once have joined hands round it.

I tried gazing at the dim-lit scene as a kind of sculptural installation, but it didn't help. The last time I'd seen trees in a cage was at an outdoor show by the sculptor Andy Goldsworthy. He'd incarcerated some dead oak trunks, stripped of bark and already as dark as coal, into a chamber of drystone walls inside a ha-ha, those picturesque ditches intended to blur the differences between the cultural and the 'natural'. So when you peered down into this man-made crevasse what you saw was not some panorama of the harmony of nature and art, but a vision of the forest cleared for the walled field, and then dumped as a *memento mori* round its edges. Not an exhibit you would fancy joining hands around, except perhaps in the hope of levitating the hapless trunks free of their prison.

What unnerved me was how *dull* I was finding the Great Yew. It had none of the panache and power and narrative fascination of deciduous trees one twentieth its age. It had no great burrs where branches had been lopped, no self-pleachings, no cryptic caverns. But as soon as I had admitted this to myself, I realised what a presumptuous reaction it was. Historical aura, visual glamour, legibility to human readers – none of these have the slightest relevance to the tree's existence, except perhaps to how it is treated.

Near the church is a pub called 'The Ewe'. Its sign is a sheep's head and yew berries rampant, a reminder that the Great Yew is now firmly on the tourist trail. I walked towards it along a series of paving stones spelling out which other beings might have passed by the 'the oldest resident in Fortingall': 'Stone Age Man … Picts … Wolves … Warriors … Roman Legions … Worshippers through the ages … And YOU'. It's another ghastly pun, but true. This commodifying of ancient organisms – only ancient by comparison with our own brief span – is about us. The conundrums of ancient yews – Were they sacred totems, planted by Neolithics? Were churches sited where yew trees already grew? – are seductive puzzles, but are really more about our social preoccupations than the life of the yew. At Fortingall it is as if we can't see the tree except in our own attachments to it; as if we hope its origins, laid bare, might reveal lost human sensitivities or beliefs. The tree itself, *for* itself, recedes. It already resembles an inanimate standing stone, confined in a space defined by us and not far off being the next piece of paving in the tourist trail.

Veteran trees began to be specially noticed in Europe during the expansive days of the late seventeenth century. They were ingredients of an increasingly valued nature as well as bits of real estate. It helped that they were rooted to the spot, that they became not just more distinctive with age but positively monumental. Big old trees consolidated place and the long span of history in a way that was rivalled only by big old houses. 'What can be more pleasant,' wrote agricultural improver John Worlidge, in 1699, 'than to have the bounds and limits of your own property preserved and continued from age to age by the testimony of such living and growing witnesses.' The big trees were items of property themselves, part of what would come to be called 'heritage'. They were often given human names. There is Wesley's Beech, Newton's Apple, any number of King's Oaks. I was once introduced to an 800-year-old oak in Dorset which was called Billy Wilkins, as if it were a vegetable scion of the local landowning family. But there aren't, to my knowledge, any yews named after humans. Even the Cumbrian trees in Wordsworth's famous poem 'Yew-Trees' are always referred to as the 'fraternal Four of Borrowdale'. Yews' brooding presence and toxic foliage didn't make them popular candidates as status symbols or the ornaments of pleasure grounds. The realisation that a few big ones, often now in the privileged environment of churchyards, had survived for millennia dawned during the heyday of antiquarianism in the mid eighteenth century. It was the beginning of a familiar process, by which a wild organism is progressively denatured. There have been claims and counterclaims about the meanings of the old yews, but they have one thing in common: the yew is seen as a human annex.

The Fortingall yew was first measured by the naturalist Daines Barrington in 1770, at fifty-two feet in girth. Two years later his friend

Thomas Pennant's tape showed fifty-six feet six inches – demonstrating how hard it is to measure these massy creatures precisely. The two men were correspondents of Gilbert White and had either prompted, or been prompted by, White's investigations into the tree in his own churchyard at Selborne. He was aware that the yew was a veteran, and thought it 'coeval with the church', which dates from the late twelfth century – the good Christian in him perhaps not ready for the implications of it being even older. White assembled the possible explanations about why there should be such trees in churchyards with his customary naturalist's thoroughness. He thought they might be there to provide shade for 'the most respectable parishioners'; or a screen from the wind; or to supply faux palm for Eastertide; or, because of their poisonous foliage, to keep cattle out of the churchyard; or, most likely, 'as an emblem of mortality for their funereal appearance'. The most popular myth, in its blend of patriotism and whimsy, was that they had been grown to provide wood for longbows. This overlooked two important facts: that the less brittle wood of Spanish and Italian yews was preferred for bows, and that they were carved from yew *trunks* – three or four at the most from each trunk. Result: the chosen tree disappears from the churchyard.

The popularising of yews had consequences in Fortingall. In 1833 the antiquarian Dr Patrick Neill reported an early black market in souvenirs of the Great Yew. Bits of the tree had been cut away 'by the country people, with the view of forming quechs or drinking cups and other relics, which visitors were in the habit of purchasing'. The nineteenth-century fad for Druidism effected a different kind of filching, an intellectual appropriation. Although there was no objective evidence, Druid revivalists declared that yews had been sacred to the cult, that they had been planted systematically around wells and other sacred sites, and that Christian churchyards associated with yews had been built on the sites of Druidic temples. Godfrey Higgins made the first claim that yews were Jehovah's own trees in *The Celtic Druids*, published in 1829.

A different slant on the Druidic theory was given by the geographer Vaughan Cornish in the 1940s. Cornish was a polymath whose interests ranged from the effect of waves on sea-shore formation to the aesthetics and human history of landscapes. In his classic study *The Churchyard*

Yew and Immortality (1946) he accepts that yew trees may have been sacred to early peoples in Britain, but argues that their evergreen foliage made them symbols of *immortality*, not mortality as White had thought. The Christian Church adopted them as symbols of everlasting life. He is less sure about when this practice might have started, but suggests that the original deployment of yew in churchyards across the English countryside was a custom brought here by the Normans, who had adopted the yew as a northern equivalent of the cypress.

Unlike earlier writers, Cornish had done his fieldwork. He wrote to every diocese in the country about their yews, and travelled to many of their locations to map their positions and alignments. The results seemed to back up his theory. Most of the very old yews appeared to lie in territories in southern England and Wales affected by the spate of church building that followed the Norman Conquest. And their positioning with respect to the church showed a surprising uniformity. The vast majority were on the south side, close to the door used by funeral processions. The coffin would pass by the yew, or sometimes between a pair of them. What Cornish refused to accept was that the individual trees themselves were especially old, or that they might be living survivors of pre-Christian religious sites. The idea of a 2,000-year-old tree, predating not just church buildings but one of the spiritual founders of Western civilisation, was still thought to border on blasphemy, demeaning to both Christianity and civilisation. The Keeper of the Department of Botany in Cardiff told Cornish categorically that 'there is no proof that any now standing date back to the time of the Druids, and it is quite unlikely that they do'. As for the Fortingall yew, there was a simple explanation given to Cornish by the Director of Kew Gardens, the distinguished Dr Edward Salisbury. The yew was not one tree, but two or even several, fused together. It was a common phenomenon, and 'the barks at the point of fusion may be completely obliterated with age, and the wood of two trunks may appear as one' (an interpretation since shown up as false by DNA analysis of different parts of the trunk). So that was it. The yew had been thoroughly explained, neatly compacted both physically and historically.

And there the matter might have ended, with all the veteran yews tidied away as pious bouquets planted at a church's founding, but for the intervention of Allen Meredith. It would be hard to imagine a man less like the respectable and academic Cornish. Meredith had a touch of the Celtic hedge preacher about him. He'd left school at fifteen with no qualifications, served in the Royal Green Jackets, lived rough for a while, got in trouble with the law. Then, in the mid 1970s, he had the series of mystical dreams with which I began this story. For the next decade he cycled about Britain, searching for the yew's 'secret'. It became an obsession, and like Cornish, Meredith did his fieldwork meticulously. He visited almost all the surviving ancient trees, measured them, probed deeper than anyone previously into historical archives, had more revelatory dreams. And he became convinced that the conventional wisdom about their ages was wrong by several orders of magnitude. He drew up a list of some 500 yews he believed were more than 1,000 years old. At Ankerwycke near Windsor, there is a yew of thirty-one-feet girth which, he argued, was the tree under which the Magna Carta was sealed; in Crowhurst, Surrey, a thirty-five footer, which he put at over 2,000 years old; and in Discoed in Wales, a tree much better preserved than the Fortingall yew, but possibly also over 5,000 years old.

These ages can make you giddy, and incredulous. They far exceeded those of any known oak trees, or indeed of most other species on the planet. And they were made problematic by the fact that old yews, at this time, were regarded as almost impossible to date. Most trees go through three distinct stages of growth. For their first fifty to a hundred years they grow comparatively quickly, and the new wood builds up in a series of wide annual rings in the trunk. In middle age (100–500 years) the size of the crown stabilises, the annual increment of new wood remains constant, and the corresponding rings become thinner and more uniform. In old age, the tree may actually shrink as branches fall off or die back, the production of new wood declines, and annual rings become very thin.

But yews are an exception. For their first 400 or 500 years they grow normally, if slowly, and can be accurately dated by removing a thin core from the trunk and counting the annual rings. But at any stage after this they can spring into a pattern of new growth quite unlike any other temperate species. They are not the only tree to become hollow in old age, but one of very few to use this as an opportunity to spring into youthful generative mode again. An old yew can begin bewildering growth spurts in new dimensions. It builds buttresses round the remaining trunk, regenerates new shoots round areas of even catastrophic damage, sends up new trunklets from branches which are low enough to lie on the ground and put down roots. Most dramatically, it can plunge aerial roots down into the hollow centre. The possible consequences of this were first reported in 1837, by J. E. Bowman, one of the most level headed of early yew investigators. Peering into the catacombs of wood inside a hollow tree in the churchyard at Mamhilad, near Pontypool, he discovered that

> in the centre of the original tree, is seen another, and apparently detached, yew, several feet in diameter, covered with bark, and in a state of vigorous growth: it is, in fact, of itself a great tree, and overtops the old one. On examination, however, it is found to be united behind, and also at some distance from the ground, by two great contorted arms, one on each side, to the inner wall of its decaying parent.

All orthodox approaches to ageing the trees falter when they arrive at an organic mass that is constantly reinventing itself and not behaving like a senior citizen at all. Even radiocarbon dating, which relies on the fact that all one-time living matter contains the C_{14} isotope, which decays at a constant rate (its half life is about 5,730 years), is useless with a trunk in which the early age ranges of wood are missing.

But the weight of Meredith's circumstantial evidence eventually led the authority on ancient trees John White to work out a tentative – though far from definitive – way of dating the old trees. It involved a formula relating the density of the rings in the outer wood to their distance from the centre of the tree. It is not a calculation for amateurs (one stage involves the calculation $[dbh/2]^2 \times \pi$) but it seemed to work quite well when tested on trees whose age was known from documentary records. When it was applied to more ancient specimens, it suggested that Meredith's informed guesses were roughly right. The big trees' ages were more like 2,500 to 3,500 years, rather than 5,000, but it still meant they were older than the churches, sometimes by a huge margin. This left matters in a teasingly unresolved state, with the various theories now equally implausible. Could early church architects have so precisely orientated their buildings that they accommodated a large lump of still-growing timber just by the funeral entry door? Is the Celtic origin of yew-bound churches credible, given that no archaeological evidence of underlying worship sites has ever been found? Is it probable that Neolithic peoples were deliberately planting a tree that grew of its own accord in the countryside around? Isn't it more likely that – if the yew really was sacred to them – they created their holy places next to existing wild trees?

I doubt there will ever be definitive answers to these questions. For many under the spell of *Taxus* this is not just a search for the history of the yew, but for a kind of spiritual genealogy, a quest for Avalon, for the deciphering of a symbol that shows how we left the path of natural religion.

In the meantime, the trees themselves continue to be just trees, and to demonstrate that their seeming immortality is neither certain nor mystical. The Selborne yew was blown down, aged approximately 1,500 years, in the great gale of 25 January 1990. The vicar's awestruck description has itself become a piece of local lore. 'The massive trunk lay shattered across the church path,' he wrote in the parish magazine, 'and a disc of soil and roots stood vertically above a wide crater. The bench around the trunk was still in place, looking like a forgotten ornament on a Christmas tree. A stormy sea of twisted boughs and dark foliage covering the churchyard was pierced here and there by a white tombstone like a sinking ship.' Also by bits of the skeletons of people whose graves had been overtaken by the roots.

After the devastations wrought by the previous great storm of 1987, the locals found the tumbling of one of the South Country's most celebrated trees hard to bear, and a rescue attempt was mounted. A team of aboricultural students sawed off the top branches and winched the trunk back into the vertical. The children from the local primary school, led by the vicar, linked hands round the risen bole to pray for its survival. It seemed to work. Shortly afterwards, unsettled by all the activity in the ground around it, an underground water main burst, and bathed the yew's roots in municipal water for the next thirty-six hours. Alas, it proved too much, and the roots were waterlogged. After a few months during which it put out a few wispy new shoots, Selborne's totem pole expired.

But the tree, echoing its species' seemingly unquenchable ability to biologically regenerate itself, refused to go away. The village planted a cutting from it just a dozen yards away from its fallow hulk – which itself lives on, by proxy at least. Its hollow shell has been colonised by young hazel and foxgloves, and a honeysuckle has wound its way up the truncated trunk. Bits of the pagan timber found their way *into* the church, as a hanging cross and an altar screen. People from all over Britain came and gathered fragmentary mementoes of the tree under which they'd picnicked, slept, made vows of undying love. I have a small section of a branch myself, now a bookend on a library shelf that holds Gilbert White's description of its parent.

In the afternoon of my visit to Fortingall, I walked from The Ewe back to the yew. Its new shoots were foxy with pollen-heavy male flowers. It looked as if it could easily live another few thousand years, but only by becoming a kind of low hedge or a rockery plant, and abandoning the energy-expensive business of keeping a trunk alive. Beings in the natural world – trees especially – don't often cling to individuality in the way we'd like them to. Their boundaries become amorphous, absorbing and joining with other living forms. The oldest living organisms in the world are probably the subterranean mycorrhiza of ancient forest fungi. They've been there since the woods sprang up, tens of thousands of years ago, and live in an intimate partnership with the tree roots, without which neither could survive. The root and fungal tissues are as imbricated as if they were a single organism. The tree supplies the fungus with sugars, the fungus filters minerals from the soil into the tree's roots. Sometimes these fungal systems stretch throughout a wood as an unbroken network of subterranean tissue, entwined with the roots

Woodcut of the Fortingall yew in 1822. The tree's hollowing out into two distinct trunks is clearly visible.

of most of the trees in the forest, an immense feeding and communication cooperative which may weigh hundreds of tons.

Individual trees, extended by suckering into genetically identical clusters, can also reach vast dimensions. A clone of 47,000 quaking aspens in Fishlake Forest in Utah – known as Pando, the 'Trembling Giant' – is, at approximately 80,000 years, probably the oldest mass of connected tree tissue, and weighs in at 6,600 tons. Another famous clonal 'tree' is the 'King's Holly' (not a holly, but a member of the protea family, *Lomatia tasmanica*) which grows in south-west Tasmania. It has flowers but never fruits, and survives purely by vegetative reproduction. Essentially, it takes cuttings of itself. When a branch falls it puts down new roots, establishing a new plant that is genetically identical to its parent. The lifespan of the individual trunks or groups of trunks is about 300 years, but radiocarbon dating suggests the whole organism is at least 43,600 years old, and the surviving 500 clonal groups are strung out in a shifting colony more than half a mile long.

In a different culture the Great Yew might have been able to expand by a similar process. On its far side I discovered a stone in the wall dedicated to a nineteenth-century incumbent of the church. A vigorous branch of the tree was beginning to overshadow it, and would doubtless soon be snipped back to the civic respectfulness of the fence line. One of the mechanisms by which yews can extend their life is analogous to the *Lomatia*'s self-cloning. The drooping branches, if allowed to grow long enough, can root where they touch the ground, and send up new trunks, which in turn form new colonising branches. The mother tree survives by expansion of the limits of its self. It occurred to me that, without its cage, the Great Yew might by now have loped as far as the door of The Ewe, in a thin green line of new incarnations.

A while later I found a churchyard yew that had been allowed to do this. In the north Wales village of Llangernyw (certainly named after its mother tree – *yw* is the Welsh-language root for the tree) the tiny church of St Digain's sports a huge and unfenced yew. It consists of four huge trunks spread out like a fistful of spillikins. It is hard to read let alone measure the tree's girth. I made it roughly forty-five feet. The official figure is forty-one feet. The age, perhaps optimistically, is put at circa 4,000 years. But the trunks and branches have extended unfettered over most of the churchyard, shading and brushing graves old and new. One branch is snaking towards a patch of bramble, and has made solid contact with the ground. In a decade or so it should have put down roots, and a new clonal offspring of the Llangernyw yew will sprout up twenty yards from the mother tree.

4

The Rorschach Tree: Baobab

THE FIRST TREES TO BE REGARDED as prodigiously ancient by European explorers were the baobabs of Africa. Baobabs evolved on Madagascar, which was cut off from mainland Africa more than 100 million years ago, and became a cauldron for the development of bizarre organisms. Ninety per cent of the island's plants and animals grow nowhere else on earth – lemurs most famously, but also three-quarters of its 850 species of orchid. There are six aboriginal species of baobab here and all are adapted to living in the parched soils of the Madagascan savanna. They have waxy white flowers which are pollinated by moths, bats and even bush babies, which have been spotted eating the petals and playing with the powder-puff stamens. To conserve water they have evolved dwarfed crowns, short branches and leaves which drop early in the dry season. When the trees are young, these flat, foreshortened topknots look like roots. A local myth explains this by suggesting that the primordial baobab was too beautiful for its own good. The gods turned it topsy-turvy as a punishment for vanity, and for good measure endowed it with cumbersome portliness.

But what chiefly keeps baobabs alive in the drought months are the paunches and elephantine buttresses that develop on mature trees. Their trunkwood is as soft and absorbent as balsa, and becomes a living cistern, capable of storing thousands of gallons of water. What is uncanny is how often baobabs resemble human water vessels – not just vaguely, but in the exact detail of accommodating barrel and narrowing neck, as if there are physical imperatives underlying the design of all liquid containers. Early Africans' own crafted vessels may not have needed the inspiration of baobabs, but it is easy to see how both species reached common solutions to the challenge of holding in a volatile, mobile material. There are baobabs reminiscent of pitchers, chamber pots, petrol cans and superior magnums of wine. At Ifaty on Madagascar, a specimen of *Adansonia za* is a gigantic three-dimensional cartoon of a canteen teapot, complete with an upwardly pointing spout. I suspect that Vermeer, whose paintings of interiors are adorned with the fluid curves of jars and jugs echoing those of their human users, would have been intrigued by the baobab tree, the vegetable pitcher and paunch.

At some point in the last 10 million years the seeds of one of the ancestral baobab species, which are cosseted in large buoyant pods, floated out across the Mozambique Channel and fetched up on the East African mainland. The seeds germinated, and the resulting trees evolved over millennia into a seventh species. *Adansonia digitata* proved to be the most adaptable and successful of the genus, and soon spread across the continent, eventually helped on its way by local people, who found it an inventive, accommodating and adaptable companion. In 1832, in the early stages of his voyage on the *Beagle*, Charles Darwin was shown a great baobab on the Cape Verde islands, 300 miles west of mainland Africa. It was reputed to be 6,000 years old, and Darwin carved his initials on it, though can't have believed this figure. Old trees like this often become hollow, at which point they may be dragooned into service as village reservoirs – a water tank inside a watery shell.

It was in mainland Africa that the baobab met its mammalian doppelgänger. Elephants (absent from Madagascar) homed in on these suggestively pachydermic invaders. They attacked them with a ferocity that seemed to go beyond the simple satisfaction of big appetites. They trashed them. They tore off whole branches, devoured the leaves, stripped the bark entirely from the lower parts of the trunk to reach the moisture underneath, and often knocked the smallest trees flat. However, the baobab had already evolved adaptations to rough treatment in the bush fires of Madagascar. When damaged or stripped, by whatever agency, the bark grows back, just as it does on stripped or burnt cork oaks. The fallen trees simply continue growing where they lie, building wooden boulders out of their wrecked trunks, pushing up new columns, snaking out new limbs parallel with the ground. One famous heap of a tree close to the Limpopo in South Africa is known locally as 'Slurpie' – an affectionate reference to the baobab's complex relationship with elephants, as both victim and mimic. It's an abbreviation of *Olifants-lurpboom*, meaning 'Elephant Trunk Tree'.

This plasticity of form is the most remarkable feature of baobabs; they are shape shifters. They can swell, shrink, curl, explode, creep about. They can begin their lives as Palladian columns, be blown down, set on fire, and then regenerate from the wreckage as a coil of snakes or a lava stream or a cave mouth. Human strangers who have witnessed their protean powers have been inspired to protean dreams.

In 1749 the young French naturalist and traveller Michel Adanson, away on an exploration of the coast of Senegal, canoed out to the island of Sor. He had planned to hunt antelope, but was stopped in his tracks by a different kind of trophy, immense, immobile, uncatchable. 'I laid aside all thoughts of sport,' he wrote in *Histoire naturelle du Sénégal*, 'as soon as I perceived a tree of prodigious thickness, which drew my whole attention ... I extended my arms, as wide as I possibly could, thirteen times, before I embraced its circumference; and for greater exactness, I measured it round with a packthread, and found it to be sixty-five feet.' It was a baobab, and Adanson was transfixed by its mass and gravitas. He later found trees with girths over seventy-five feet, and concluded that 'as Africa may boast of producing the largest of animals, viz. the ostrich and the elephant, so may it be said, not to degenerate with regard to vegetables, since it gives birth to calabash trees, that are immensely larger than any other tree now existing, at least that we know of'. He soon began to suspect that they were also the oldest trees on earth. Deep in the interior of Senegal he found baobab trunks inscribed with the names of European settlers from the fifteenth and sixteenth centuries, the letters of the signatures still legible, and barely stretched beyond the original knife marks. In the compelling fixity of

'Baobab Avenue', Morondava, Madagascar.

these graffiti, which seemed immune to the ravages of time, Adanson thought he saw evidence of organisms whose laborious growth had begun maybe 5,000 years ago, before the Flood.

This was a heretical thought for the time, and outraged a later Africa hand, the staunchly Christian Dr David Livingstone. In *Missionary Travels* (1857) Livingstone wrote: 'Though it possesses amazing vitality, it is difficult to believe that this great baby-looking bulb or tree is as old as the pyramids.' Making a rough estimate of the annual growth rings (they are indistinct in the species), he reckoned that a tree one hundred feet round would be only 1,400 years old. Modern measurements, using radioactive carbon dating, have confirmed that baobabs rarely reach one fifth of the age Adanson imagined. Their awesome bulk had persuaded him to see them as primeval.

So here was a classic Enlightenment encounter between plant and imagination. A wide-eyed, twenty-two-year-old naturalist dowsing ancient texts carved in the living body of a tree. Adanson was a precocious and irreverent young man. Thirty years later he would compile 150 volumes covering the entire array of creation as it was then known, and see them politely ignored by publishers. His ambition was to find, or make, order in a natural world he saw as 'a confused mingling of beings that seem to have been brought together by chance'. In the encrypted, living monoliths of Senegal – the genus was subsequently dubbed *Adansonia* in his honour – he had glimpsed the chaotic enormity of biological space and time, and maybe his own intellectual fate. What the baobab's capricious inventiveness represented would always triumph, however much he tried to regiment and unmingle it.

It's no wonder that the biggest and oldest trees have become *places*. Their eccentric bulk makes them obvious landmarks, sites for village meetings and religious rituals. Recently their tendency to become hollow has been exploited for more mundane utilities – bus shelters, storerooms, occasional refuse dumps. One in South Africa has been turned into a rather smart pub, with a lantern-lit bar and naturally water-cooled cellar inside. In Senegal, villagers once buried their dead in the hollow trunks, and there is no disrespect seen in these contrasting sacred and secular uses. In nearby Burkina Faso, the death of a baobab is itself commemorated by a full-dress communal wake. The trees aren't regarded as gods or wreathed by unbreachable taboos, though they are seen as providing resting places for ancestral spirits. When a number of revered baobabs in Zambia were due to be flooded by the creation of the Kariba dam, the locals 'evacuated' their resident spirits by lopping branches from the doomed trees and attaching them to other baobabs outside the flood zone.

Ancient baobabs are looked on as arboreal village elders. When they die, they are mourned like dearly departed citizens. The West African writer and storyteller Seydou Drame has described just such a wake for the Baobab of Kassakongo: 'One day its leaves failed to grow again. Still upright, the old wooden elephant had given in to death, and so it was the whole village prepared for its funeral ... The chief tells the story of the tree's life as if he were talking about an old man who had just died. "He chose to take up residence in Kassakongo. The villagers took his presence as a blessing from God."'

The ever-buoyant pods of *Adansonia digitata* eventually bobbed (or were boated) all the way across the Indian Ocean to Australia, where their descendants evolved into a very similar eighth species, *A. gregorii*. In the Herbarium at Kew there is a virtual chart of its colonisation of north-central Australia in the specimen sheets of nineteenth-century botanists, as they themselves edged through the unexplored wilderness. The difficulties involved in mounting thick twigs and fragments of fruit on flat sheets in the field (and often in the dark: some of the sheets are black with soot from oil lamps) hasn't stopped the insistent diversity of this species showing through. Different specimens from isolated valleys in Kimberley have distinctly different leaf shapes and patterns. This diversity – and the presence of what appear to be images of baobabs in the remarkable and atypical 'Bradshaw' rock art of Kimberley (dated to about 15,000 to 18,000 BC) – has encouraged some scientists to speculate that various baobabs were brought to Australia deliberately, during transoceanic human migrations from Africa 60,000 years ago.

The 'Prison Boab', Derby Harbour, Western Australia, believed to have held Aboriginal captives.

As is always the case when plant migration and cultural origins are entangled, this is a highly contentious theory, which runs counter to the orthodox view that Aboriginal culture sprang from a single colonisation by South East Asians maybe 50,000 years ago. Genetic fingerprinting of the Kimberley baobabs may soon establish the date of their arrival and how closely they're related to African trees. But it is unlikely to resolve whether they floated over of their own accord or were carried in reed boats by Palaeolithic San explorers.

Either way, *A. gregorii* has been thoroughly naturalised, to the extent that it has its own Australian vernacular tag of 'boab' (and just occasionally 'boob'). One distinctive specimen has an especially sad and strange history. The Prison Boab near Derby Harbour in Western Australia has grown into the shape of the pots in which missionaries were supposedly cooked by cannibals. Legends testifying to a rather different history weave about it. It may have been an Aboriginal burial site, then, in a dark reversal consequent on European colonisation, a place where Aboriginal prisoners, chained together, were lodged during the last stage of their journey to the local courthouse. When the explorer Herbert Basedow combed through the litter inside the trunk in 1916, he found bleached human bones and a skull with a bullet hole. The outside of the tree is riddled with graffiti and autographs, a globe-shaped rune whose origins and meaning are unlikely ever to be deciphered.

5

The Big Trees: Sequoias

IN THE NEW WORLD no one had any doubts about the potential age of trees from the moment that a United Water Company ditch digger called Augustus T. Dowd came face to face with one of California's 'Big Trees' in 1852. In a strange echo of Adanson's meeting with the baobab, he'd been out hunting, and had blundered, face to trunk, into something incomprehensively more immense than the grizzly he was tracking. It was fifty feet in girth and its top was barely visible. Dowd was the first white man to see one of the mammoth sequoias of the Yosemite Valley. But thoughts of transcendent longevity or American heritage were not on the minds of the community of dirt-poor loggers and miners who inhabited the shanty settlements in the valley. They saw booty, wooden gold. And with frontiersmen's hunting reflexes, they treated the trees, in Simon Schama's words, as 'trophy: [to be] skinned, mounted and displayed for bragging and cash'.

In the summer of 1854, maybe motivated by the success of Phineas Barnum's freak shows, another ex-miner, George Gale, thought he could cash in on the monstrousness of the sequoias. He picked the biggest he could find, ninety feet in girth and known as the Mother of the Forest, and in an extraordinary feat of surgical lumberjacking, stripped the bark off to a height of 116 feet. He shipped the pieces to New York, where they were reassembled as a hollow cylinder and exhibited as a vegetable wonder, the death mask of a living arboreal giant. But the public thought it was hoax, stitched out of fragments of many trees, and Gale's dreams turned to leaf litter.

Meanwhile, back in Calaveras Grove, where the first Big Trees had been discovered, no such urbane cynicism was on show. With no legislation to control the exploitation of the sequoias, something close to an arboreal amusement park was developing. The first party of tourists arrived in 1855, and found the same kind of diversions – transmuted out of immemorial American lumber – as were available in downtown San Francisco. Each giant tree took five men three weeks to fell, but one had already been planed flat to accommodate a two-lane bowling alley. The stump of another had been smoothed down to make a dance-floor on which, the entrepreneur James Mason Hutchings records, 'on 4 July, thirty-two persons were engaged in dancing four sets of cotillion at one time, without suffering any inconvenience whatever'.

There wasn't, as yet, much stirring of transcendent wonder about the trees, or suggestions that they were objects of sublime natural grandeur or emblems of their mother country. But scientists who had heard the stories realised the backwoodsmen had discovered something special. English botanists, feeling the Yosemite giant's height and bearing merited some kind of heroic title, dubbed it *Wellingtonia gigantea* after the grand Duke. But the French botanist Joseph Decaisne, thinking it was related to the coast redwood, *Sequoia sempervirens*, decided that *Sequoia gigantea* would be more appropriate. Asa Gray, founder of Harvard's Botanical Garden, agreed. As it turned out, the two species were not closely related, but the author of the official state *Yosemite Handbook* was happy to pass over the vagueness of the kinship links. 'It is to the happy accident of the generic agreement of the Big Tree with the redwood,' he wrote in 1868, 'that we owe it that we are *not* now obliged to call the largest and most interesting tree in America after an *English* military hero.' It was an early step in the elevation of the redwoods to the status of national symbols and affirmations of America's manifest destiny as a New Eden. These giants seemed primordial, and their first witnesses thought they were oldest living things on earth. The otherwise blasé traveller Horace Greeley was stopped short by the thought that they were standing 'when David danced before the Ark'. Another writer, Charles Fenno Hoffman, made explicit comparisons between old Europe, cradle of barbarism and feudal oppression, and

The stump of a giant sequoia in Calaveras Grove, California, 1862,
polished down to become a dance-floor.

'the deep [American] forests which the eye of God alone has pervaded and where Nature in her unviolated sanctuary has for ages laid her fruits and flowers on His altar'.

In the heart of the Civil War, Abraham Lincoln sensed that the Big Trees were a revelation of the uniqueness of the Republic and a symbol that transcended sectional differences. On 1 July 1864 he signed a bill which granted them to the State of California 'for the benefit of the people, for their resort and recreation, to hold them inalienable for all time'. But elsewhere, big sequoias continued to be felled for timber. Frontiersmen saw no contradiction between regarding a piece of nature as magnificent and appropriating it as a trophy. It was their slice of the Land of the Free.

Ten years after Lincoln's grant, Albert Bierstadt painted his famous portrait *Giant Redwood Trees of California*. The immense 'cathedral grove' of trees is bathed in red light, more like the forests of the coast redwood. But it gathers together all the iconography which had accumulated around the giant redwoods – the immense height of the trunks, which have scarcely begun to taper by the time they reach the top of the canvas, their seeming immunity to toppling or disease, their setting in a primeval landscape just touched by mist. He also includes three Native Americans, who appear to have their home in a hollow at the foot of the foremost tree. As Schama writes, 'Bierstadt's painting is sylvandomestic: the ancient residence of the *most* indigenous Americans.'

In 1901 another American president acted on the sequoia's behalf. Just after his election that year Theodore Roosevelt read *The Mountains of California* by the Scottish-born writer and conservationist John Muir. A lover of the outdoors himself, he was moved by Muir's lyrical description and passionate defence of the wild landscapes of the west. He wrote to the author suggesting that they might meet in Yosemite to discuss its future, and by implication the future of all America's wildernesses. The outcome, two years later, was a three-day camping trip to the valley, and a night spent in conversation round a camp fire in Mariposa Grove, the site of some 500 of the biggest sequoias. Muir urged the president to take the whole of Yosemite, including Mariposa, into federal custodianship.

His plea worked, and three years later Roosevelt signed the Yosemite Recession Bill which instituted the entire valley as a protected national park. In the course of his presidency he legislated for five more national parks and more than 200 protected wild areas.

There was an ironic sequel to the elevation of the sequoias to the status of a national monument, which echoes the ambivalence of their discovery years, and maybe Roosevelt's ambivalence as a man who saved trees but hunted bears. The sequoias began to acquire names. They weren't dubbed after their forms or the character of their locations, but given political titles, as if their natural gravitas annexed them to the governing elite. There are two 'Lincolns' and one 'Senate'. The biggest, in the sense of heaviest – it weighs in at 1,500 tons – was honoured as 'General Sherman', the most brutal of the Union commanders in the Civil War, prefiguring the use of his name for a World War II tank. And in a final act of hubris, the politicians at Washington conscripted the entire population of Sierra Nevada sequoias into their men's club, and called them simply 'the House'.

6

Methuselahs: Bristlecones and Date Palms

BRISTLECONE PINES, by common consent the oldest trees on earth, seem not so much old as stone dead. The snags of contorted wood that grow in the high mountains of south-western North America have the look of fossil trees, bleached to the colour and texture of the shattered dolomite rock round their roots.

There are three species of bristlecone (named from the prickles on the female cones) growing in the arid highlands between Utah and New Mexico, at altitudes between 5,600 and 11,000 feet. In this extreme habitat of long, cold winters, low rainfall and high winds, bristlecones have few competitors and are supremely well adapted. Their wood is dense and resinous, and resistant to invasion by wood-boring insects and fungi. The roots are shallow and spread wide, to give support against the wind and rapid access to surface water. The needles are coated with wax to reduce moisture loss, and can hang on the tree for up to forty years. But the bristlecones' major strategy for survival into extreme old age is, counter-intuitively, to take itself very close to death. As an already ancient tree edges into venerability much of the wood dies back, often leaving just a wisp of living tissue connecting the roots to a handful of twigs. Effectively it has become torpid, embalmed, reducing its growth – and therefore its needs – to almost nothing. The stresses of old age in bristlecones are purely climatic, not metabolic. The heartwood is so dense and desiccated that instead of rotting it is eroded like stone, by frost and gale-driven rock dust.

The species capable of reaching the greatest age is *Pinus longaeva*, and in 1957 Edmund Schulman discovered a specimen in the White Mountains of Inyo County, California, which he was able to give an age of exactly 4,846 years. By drilling an unbroken core from the heartwood he'd counted the annual rings through a microscope. At that moment it was the oldest accurately dated tree in the world, and was promptly dubbed Methuselah. (The Fortingall yew may be older but that can never be proven.) A few years later an even older tree was found nearby by a graduate geography student from the University of Carolina. Donald Rusk Currey was studying how clues to climate and even year-by-year weather are preserved in tree rings (warm wet summers produce wide rings, for instance) and bored a routine core from a tree which on his schedule was labelled as WPB-114. What he didn't know was that it was one of the most celebrated bristlecones, with its own name – Prometheus – given by local tree enthusiasts in the early 1950s. Unfortunately Currey's specialised drill got stuck in the trunk. Without it he couldn't continue his research project. So with the bravura of a frontier lumberjack, he simply cut the tree down. The story goes that he took a slice of the trunk back to his motel and sat outside in the sun to count the annual rings. There were 4,844, but he thought the tree might be older than this, as his section wasn't from the base of the tree. He imagined it might exceed 5,000 years. Whatever its exact age, at the moment of its summary execution Prometheus then succeeded Methuselah as the oldest known tree in the world. (Both it and the still surviving Methuselah were beaten in this not wholly edifying league table by the discovery in 2013 of a bristlecone from the same area with 5,065 annual rings.)

Death from dendrological drill is unlikely to happen to any of the now much more securely protected ancient bristlecones. But their future looks uncertain nonetheless. Climate change is raising the average temperature of their alpine redoubts. The prolonged frosts that killed off predatory insects are now less frequent. A new fungal disease from Asia, blister rust, has joined the list of invasive parasites that seem to have a particular taste for trees, and is attacking bristlecone saplings. *P. longaeva* seeds well, and its populations are widely scattered, so the species is not under serious threat. But Champion Trees, as they are known in the USA, attract human champions, who are interested not so much in the survival of the species as prolonging the life and exact genetic essence of the veterans. There is a whiff of arboreal eugenics here, and a nostalgic sense of heritage which believes that the best lies in the past. The ideology of veteran tree conservation – certainly in the USA – still echoes with the nineteenth-century conviction that the anciently pristine may be the God-given route to the future.

In the 1990s a Michigan nurseryman called David Milarch began a project to clone Champion Trees, including the ancient bristlecones. Uncannily, he'd been prompted by visions almost identical to those experienced by Allen Meredith before he began his research on ancient yews. Milarch had an out-of-body experience during acute liver failure, and for months afterwards had early morning visitation by 'light beings'. 'The big trees were dying, they told him, it was going to get much worse, and they had an assignment for him.' His mission resolved into one of cloning the big trees and planting them across the USA. They were a unique biological legacy, he believed, and humankind needed them. Milarch's own reasoning was disarmingly down to earth. '[T]hese are the supertrees,' he said in an interview, 'and they have stood the test of time. Until we started cloning the nation's largest and oldest trees, they were allowed to tip over, and their genes to disappear. Is that good science? If you saw the last dinosaur egg, would you pick it up and save it for study or let it disappear?' Alas, for all his good intentions, Milarch's science isn't too good either. Size and old age clearly indicate past success in surviving the 'tests of time'. But they do not predict an ability to survive the tests of the future, which may be very different. New exotic diseases and climatic anomalies are multiplying. The Champions' survival may also have resulted from a lucky combination of site and historical experience, and a clone transplanted to a new situation might have no such luck. Nor do genes wholly 'disappear' if an individual tree dies. They're retained in its relations and the offspring it may have been generating for thousands of years, and which might hold not only the genes which contribute to longevity, but others temporarily in recession (for enduring a warmer climate perhaps). The dinosaur's egg is an unfortunate comparison too, as nature invented sex (and eggs) precisely to ensure plenty of gene swapping, and broods of offspring with a wide mix of potential responses to unpredictable situations.

The seeds of old trees would be considered part of their legacy if they weren't so lacking in the monumental splendour of their parents. Many can live for great ages (though what 'live' means in the case of a dried-out husk is an interesting question), carrying potential templates not just of the particular tree that bore them, but of all its ancestors.

The longest proved dormancy for a seed is for a date discovered in the ruins of the ancient Israeli settlement of Masada. The remains of this city had lain undisturbed for nearly 2,000 years before they were excavated by archaeologists in the 1960s. In the deepest layers the diggers discovered a cache of provisions: grain, olive oil, wine, pomegranates and a generous hoard of dates from the extinct Judean palm, so perfectly preserved that scraps of fruit flesh still clung to the seeds. Four decades later, after museum workers had cleaned and catalogued the dates, someone decided to plant one. It was a shock and a thrill when the workers noticed a lone shoot sprouting from one of the pots. By 2012 it was ten feet tall, and sharing the name Methuselah with the unrelated dwarf pine halfway across the globe. Alas, this may be only a temporary resurrection. Methuselah has flowered, but proved to be a male tree. Unless a female fossil date can be found no future generations of Judean palms will follow.

Dormancy is an extraordinary phenomenon both botanically and philosophically. No one is yet sure how it works, or what exactly it means. It may be a purely physical trick, depending on the seed's coat mummifying its life processes until it is struck by just the right combination of light and temperature. Yet the seeds of some species (e.g. the field poppy) appear to be programmed to germinate far in the future, as an insurance policy. Do they contain some so-far undetected bioelectric timer? And what might be a botanical pathologist's verdict on the existential status of a dormant seed? Can it be said to be truly alive

The bleached trunk of a dead bristlecone pine, Inyo National Forest, California.
The foliaged trees beyond may be thousands of years old.

without any discernible metabolic activity? The answer given by one American seedbank botanist, Chris Walters, is: 'If seeds are alive but aren't metabolising, then maybe we need to rethink our definition of what it means to be alive.'

7

Provenance and Extinction: Wood's Cycad

IN THE CASE OF ONE INDIVIDUAL from an ancient tree lineage, cloning was the last option if the species was to survive. Only a single specimen of Wood's cycad, *Encephalartos woodii*, has ever been found in the wild, in Natal in 1895. It died as a result of repeated mutilation seventy years later while in intensive care in a government enclosure in Pretoria. Luckily a few offsets from the main stem had been despatched to a handful of botanic gardens, including Kew, before its demise. Wood's cycad is dioecious, and has male and female reproductive structures on different plants. Since no female has ever been seen, cloning from offsets is the only way Wood's cycad will survive as a true species. Its decline underscores the poignancy of extinction, but at a deep cellular level is also a heartening parable about the tactics plants employ to avoid such extreme events, and how humans can intervene to help.

The cycads are the one of oldest orders of trees. They evolved about 280 million years ago, in the wake of the Carboniferous coal forests, and flourished throughout the reign of the dinosaurs, their lush evergreen leaves providing very desirable forage. They resemble palms in their topknots of divided leaves and scaly trunks, but, as prototypes of the various kinds of plants that were to follow, they have reproductive organs which are a kind of compromise between a flower and a cone, known scientifically as strobili. The male cones of cycads can be huge. *E. woodii*'s is a yellowish cylinder which can grow up to three feet in length, its spiralling scales giving it something of the appearance of a

The fossil frond of a cycad, found in Yorkshire.
The genus flourished in the Jurassic era.

stretched pineapple. It produces an abundance of pollen, which is carried to the female strobilus – not a cone, but a large cluster of woolly leaves – by beetles and weevils. In the Jurassic era this may have been one of nature's first experiments in symbiosis between plant and insect. It was a successful experiment, and the descendants of those first pollinators would have been primed to perform the same services when flowering plants arrived on the scene.

Cycads continued to thrive during the Jurassic, and spread globally throughout the tropics and subtropics. They are still widely distributed, across South East Asia and the Pacific, the southern half of Africa and South and Central America, where the greatest diversity occurs. More than 300 different species have so far been described (new discoveries are being added all the time) and they occupy every kind of habitat, from rainforest to semi-desert. For such primitive plants, they are wonderfully resourceful, which is perhaps the reason they survived the ecological catastrophe that put paid to the dinosaurs. There are adaptations for every kind of contingency. If pollination fails they can sprout vegetatively from the base by offsets and suckers. The growing shoots are protected by fire-resistant basal leaves. They have unique 'corraloid' roots which live in symbiotic partnerships with algae able to fix atmospheric nitrogen – a trick which the bean family also achieved but 100 million years later. Their seeds are huge and surrounded by layers of nutritious husk, but have an inner seed kernel impervious to even an elephant's digestive system. The seeds of some species can float, possibly over whole oceans, like the baobab's. They also, as a family, have a strong propensity to throw up new subspecies and hybrids, adapted to quite specialised niches, providing yet another route to survival and the colonisation of new territory.

But the life of a fine-tuned, highly local species is also precarious, not suited to sudden changes in its environment. *E. woodii* was one of these vulnerable endemics. The only known wild individual was a cluster of four stems discovered in 1895 by John Medley Wood, curator of the Durban Botanic Garden, growing on a steep, south-facing bank on the edge of oNgoye forest in Natal. A basal offset was removed and sent to London's Kew Gardens in 1903, where it has grown in the Temperate

House ever since. Aware of the tree's knife-edge existence, botanists took more basal cuttings over the following decades, and sent them to botanic gardens in Durban and Ireland. By 1912 there was only one short and damaged stem of the original plant left in the wild. The priority of conserving a species in its wild habitat wasn't properly understood at the time, and four years later the Forestry Department arranged to have this last surviving specimen dug up and sent to the Government Botanist in Pretoria, where it expired half a century later. *E. woodii* has consequently been officially granted the dubious honour of 'Extinct in the Wild' status.

Currently offsets from the Durban and Kew trees are growing in other botanic gardens across the world, an example of the curious but expanding process of the democratisation of rarity. Kew's specimen, now more than a century old, will be a star feature of the New Temperate House to be opened in 2015. Meanwhile it lies huddled in a corner in the original building, shorn of its fronds, as these had grown too big for the plant to get through the door – a cutting down to size that seems to echo its past history. But Wood's cycad will never reproduce naturally unless a female is discovered, so current attempts at propagation rely chiefly on the fact that its pollen will generate fertile hybrids from its close relative, *E. natalensis*. If these offspring are repeatedly back-crossed with *E. woodii*, a progeny may eventually arise that will show roughly what a female *E. woodii* looked like in the wild, echoing the scientific dream that woolly mammoth hybrids may be brought back from the dark world of the extinct by inserting fragments of fossil DNA into elephant genes. Our sense of 'the natural' warps here. The thrill of Promethean creation, of a pathway to reparation for the damage we have done to the natural world, crashes into an ancient Frankensteinian guilt and a simple longing for a world beyond our planning. But these things will happen and we need to learn how to think ethically about them.

Worries about the individuality of ancient plants are especially acute with trees. Their age and singularity give them something of the aura of living artworks, and provenance and authenticity become issues. In America the seniority of coniferous Methuselahs is fiercely contested. In

Britain communities make claims and counterclaims about whose oak a displaced monarch or fleeing dissenter once hid in. The trees themselves are fenced and proclaimed on ornamental plaques like military heroes. Hybrids between native rarities and more vigorous intruders are despised and sometimes destroyed because, from a niggardly view of biodiversity, they are diluting the genetic purity of the original. It is as if individual trees with character and rich biographies can continue to have an existence only through the pickling (by cloning, for example) of their unique genetic identity; and that being carried forward by the eddying, unpredictable streams of reproduction, as all other organic life is, would obliterate their authentic essence. As a principle guiding our treatment of nature as a whole this, it hardly needs saying, would have stopped life on earth in its tracks. We too often forget that trees have been successfully negotiating all the processes to which we subject them – mutation, evolving adaptations to changed circumstances, cross-breeding, self-planting, regenerating – entirely of their own accord for millions of years.

There's an appealing young Wood's cycad in the Hortus Botanicus in Amsterdam. It grows, unprotected, in a small wooden tub of the sort you might use for a favourite rose by the front door. Its next-door neighbour in the small conservatory is a tiny olive tree which has been potted in an old olive oil can, a pairing whose subliminal message is that all plants are equally valuable, regardless of their human usefulness and entanglements. I like the unpretentiousness of this modest offshoot of 200 million years of history. But I've experienced the *mana* that can attach to Big Trees – the Old Ones, the survivors against the odds, the lasts of their kind. They can reverse history's telescope, and in their singularity sometimes be portals into deep time, through which you can glimpse evolution rolling backwards: the solitary tree becoming the grove, then the teeming forest, then the community of aboriginal species. You can look forward from them too, perhaps to the moment when they cease to be and all that is left, in Oliver Sacks's words, is 'the sad, compressed memory in the coal'.

Sacks – a man equally visionary as both botanist and neurologist – became obsessed with ancient species as a child, when he was taken to see dioramas of the Jurassic Age in the Natural History Museum and to the cycad beds in Kew's great Palm House. He was transfixed by their overwhelming sense of antiquity and strangeness, and had dreams of them as 'an Eden of the remote past, a magical "once" … They neither evolved nor changed, nothing ever happened in them; they were encapsulated in amber.' Later, as an adult, he understood that their peculiarity magnified his understanding of the essential dynamics of life. He saw the cycads as both 'tragic and heroic'. The coal-forest plants they were most closely related to have long vanished from the earth, and the cycads find themselves 'rare, odd, singular, anomalous, in a world of little, noisy, fast-moving animals and fast-growing, brightly coloured flowers, out of sync with their own dignified and monumental timescale'.

In the 1990s Sacks visited a string of islands in the Pacific where a strange endemic colour blindness exists that might possibly be related to regular consumption of flour made from cycad seeds. He tells of one evening on Rota, a small island east of Guam, when he sits on the beach, amongst cycad trees that come down almost to the water's edge. The sand is littered with their giant seeds, and fiddler crabs are emerging to scissor their way in for the kernels. A light wind has got up, and the strengthening waves are catching some of the seeds and drawing them into the sea. Most are washed back onto the shore, but Sacks watches one surfing the waves, beginning to edge away and perhaps on the beginning of a journey across the Pacific. There is a chance that it will pitch up on another island and germinate; and a more remote but thrilling possibility that it might eventually hybridise with another cycad species there, and extend the family's survival opportunities into the future. If a human hand helps in the fertilisation, it would seem to me no different from the lucky intervention of an enterprising weevil, simply speeding up a process that could have happened entirely naturally. Similarly, in the journey between hunter-gathering and domestication it is often difficult to tell whether it is plants or people that are driving the developments.

8

From Workhorse to Green Man: The Oak

THE OAK FAMILY was never likely to supply a candidate for the Tree of Life. Oaks are too workaday, too down to earth, too uncompromisingly *woody*. They don't aspire to symmetry or spire-like elegance, but work out their destinies through obstinate and not always pretty eccentricity. Their family name, *Quercus*, ought to be the root of 'quirky' but, perversely, seems not to be.

There are between 400 and 600 species of oak spread across the northern hemisphere, from Colombia to north-east China. The vagueness of this tally reflects not just the endless squabblings of taxonomists (by no means ended by developments in molecular biology) but also something essential about the *Quercus* family. It's opportunist, mutable, full of hybrids and intensely local varieties. There are majestic white oaks in North America and small evergreen 'ring-cups' (so called from the wide fringes on the acorns' cups) in South East Asia. In the New Forest there is a race of English oaks which spring into pale and short-lived leaf around Christmas and then again in the spring. The family repeatedly overthrows cultural and botanical preconceptions. Despite Britain's belief in a special relationship with the tree, as the nation's 'heart of oak', the oaks' botanical heartland is Mexico, where there are 160 species, 109 of which grow nowhere else.

The behaviour of a single species, the Mediterranean prickly oak, *Q. coccifera*, typifies the family's plastic qualities, and forty years after my first encounter I still have problems identifying the more outlandish specimens. It's a tree mimic artist, capable of morphing into almost any imaginable arboreal form according to its circumstances. I've seen it as a dwarf and spiny shrub in the garrigue in Provence, grazed down to no more than four inches in height but still bearing acorns, and as a stately sixty-foot tall, ten-foot round timber tree in the Lassithi Mountains in Crete. The way a single individual can effortlessly pass between these two extremes suggests a held-in-reserve adaptability that may be epigenetic, with different forms being switched on and off according to circumstance. *Q. coccifera* isn't killed by felling, fire, shade or sheep. It regrows from the stump or rootstock, and the harder it's grazed the more protectively spiny the new leaves become. Low-intensity browsing tends to push the regenerating oaklet into the shape of a classical column, slightly ruined but determinedly upright. The shoots around the base of the shrub spread out laterally until animals are unable to reach the centre shoots – which then grow upwards, allowing the oak to 'get away'. It may eventually become a mature tree with low branches, at which point agile browsers can scramble up and edge their way along the woody tightropes, munching foliage just as they do on the ground. It is an extraordinary sight, trees bearing animals like fruit, with sheaves of shoots, browsed bare of leaves, rising vertically from the main horizontal branches. Historical ecologist Oliver Rackham calls them 'goat pollards'.

The family doesn't field many record breakers. Oaks don't figure highly in the lists of the oldest, tallest, strongest or most massive Champion Trees. But one species or other has usually been around wherever northern people have lived, and their general all-round woodiness made them pillars of regional cultures. Throughout the Neolithic era oak was a crucial raw material, providing firewood, axe handles, frames for shelters. The north European oaks *Q. robur* and *Q. petraea* can be split cleanly, even with stone axes, and flat oak planks – imaginatively provocative artefacts in a world dominated by naturally curvaceous forms – paved the earliest surviving European walkways. The Sweet Track that crosses the Somerset marshes in England is surfaced with oak planks, supported by a scaffolding of ash, lime, elm, alder and oak poles, almost certainly grown in worked coppices. The planks have mortises bored in them so that they can be pegged to the framework. The wood is so well preserved that recent advances in dating trees by their growth rings has placed the cutting of the wood precisely, at between the years 3806 and 3807 BC.

Availability and durability were northern oaks' prime virtues. They could be turned into almost any kind of structure where long life and resistance to weathering were crucial. They supplied the timbers for Viking warships and Christian churches. In the village of Allouville-Bellefosse in northern France, a 1,000-year-old tree is known as the Chêne chapelle. Its hollow trunk hosts two fully functional chapels, built in 1669 and still used for Mass twice a year. The tree's unlikely progression to the status of sacred architecture began when it was struck by lightning and burnt hollow. The local clergy claimed that the lightning strike was a heavenly bolt, and that the resulting hollow had a sacred purpose. During the French Revolution it became a symbol of the ancien régime and the tyranny of the Church, and a crowd attempted to burn the entire structure to the ground. But a local citizen seemed to intuit the oak's mutable character and renamed it The Temple of Reason. It became, temporarily, a symbol of the new democratic thinking, and was spared.

The roof of Westminster Hall, which contains 600 tons of wood spanning seventy-five feet without a central support, is another ecclesiastical marvel and the most remarkable wooden roof ever made. It is one of those paradoxical (or inevitable) wooden structures where the carpenters, having dismembered large numbers of trees, then reassemble them in what is essentially a supertree, a formal arrangement of trunk and branching that would not work unless it aped the structure of its motherlode. William Bryant Logan, one of the oak's recent biographers, explains:

> Each member of a roof truss – the triangular or spire assembly of rafters – is a force made visible. Gravity flows down the roof rafters and pushes at the walls on which the roof sits. The tie beam resists, holding the two pieces in tension. Above the tie beam, the carpenter places a collar – a smaller and higher version of the tie beam – to siphon off some of gravity's pull. Beneath the collar, a post or a pair of arched braces would

let the forces cascade down to the centre of the tie beam. Beneath the tie beam, more arched braces led out to a lower level of the walls, letting the forces run out into the ground.

But in Ely Cathedral – the 'Ship of the Fens' – whose soaring and labyrinthine interior resembles a carved simulacrum of an oak forest, the tree's idiosyncrasies won out over the architect's ambitions. The inner timber tower rests on sixteen struts which are meant to be forty feet long and over a foot square, but, as Rackham remarks, 'with all England to draw on for the timber the carpenter evidently had to make do with trees that did not quite meet this specification'. The struts taper rapidly at the top, where they had formed the crowns of their original trees; in six cases the design was altered to use trees that, even so, were not quite long enough to reach up to the spectacular octagonal lantern in the roof.

It wasn't just the obvious strength of oak that made it a symbol of nationhood in Britain. If North Americans glimpsed their country's pioneering spirit and unsullied, heaven-blessed landscapes in the soaring of the giant redwoods, the British saw in their sturdy, tenacious oaks, ingrained with Old World history, something of their own pugnacious and decidedly un-transcendental character. After David Garrick had published his patriotic sea shanty 'Heart of oak are our ships/ Heart of oak are our men ...', the naval historian John Charnock floated the audaciously nationalistic idea that only British soil could grow oakwood suitable for ships which aspired to the spirit of British nationhood:

> It is a striking but well-known fact that the oak of other countries, though lying under precisely the same latitude as Britain, has been invariably found less serviceable than that of the latter, as though Nature herself, were it possible to indulge so romantic an idea, had forbad that the national character of a British ship should be suffered to undergo a species of degradation by being built of materials not indigenous to it ...

The diversity of oak-tree form has been reflected in the ingenious and frugal uses found for its various component parts. The tannin-rich bark was used to soften hides in manufacture of leather. The leaf galls (produced by wasp larvae) were the source of an intensely dark ink, used by Leonardo da Vinci in his drawings. Acorns (*balanos* in Greek) crop up as phallic symbols in classical sculpture, and modern medicine has returned the compliment by naming inflammation of the penis *balanitis*. In Spain evergreen oaks are the mainstays of an entire rural economy. The cork oaks, *Quercus suber*, have their bark stripped every nine years. The acorns provide food for the local pigs. The branches, and those of neighbouring holm or live oaks, *Q. ilex*, are cut to increase the acorn crop, and the prunings converted to charcoal, bound for the barbecue market. The roast acorns of both species are sweet enough to be a popular Iberian bar snack for humans, sold as *bellotas*. The acorns of the South East Asian *kunugi*, a variety of *Q. acutissima*, are also still traded commercially as a source of starch. Modern Brits could never have granted a staple role for the intensely bitter acorns of their native oaks, but during the domestic austerities of World War II, the Ministry of Food suggested making a substitute coffee from them. They had to be chopped, roasted, ground up and roasted again. The resulting beverage was acrid and caffeine free, not a product a nation under siege was likely to take to its heart, or palate. (Another indigenous plant came to the rescue when roasted chicory roots were conscripted instead.)

The quercophilic William Bryant Logan was struck by this widespread use of acorns as human food. Researching for his book *Oak: The Frame of Civilisation* in 2004, he came across a map of 'World Oak Distribution'. He was astonished to find that, in his reading, 'the distribution of oak trees is coterminous with the locations of the settled civilisations of Asia, Europe and North America'. Japanese and Korean customs of eating *kunugi* acorns, for instance, were echoed on the other side of the Pacific, where indigenous North American peoples used white- and live-oak acorns as a staple source of carbohydrate. It ought to have been no great surprise. Most humans flourish in the same climatic and environmental conditions as most oak species. But Logan was sufficiently moved by the coincidence to frame a radical new theory about the origins of civilisation. He dumps the conventional idea that hunter-gatherers and early pastoralists evolved gradually into cereal farmers

Boars feeding on acorns, in the ancient practice known in Europe as 'pannage'. This illustration is from a late medieval Latin translation of an Arab medical treatise.

through close contact with the wild grasses eaten by their half-wild stock. He argues instead for an intermediate stage in which acorns – communally gathered and stored over winter – became the exclusive, worldwide prototype for centralised agriculture. Oaks and human settlements were not 'coterminous' by coincidence or shared habitat preferences, but because people had deliberately moved to where oaks grew, in search of the staff of life.

Few archaeologists would argue with the notion that late hunter-gatherers were collecting and storing nuts and fruit, and maybe beginning inadvertently to cultivate them, whenever discarded pips and kernels germinated near settlements. But the range of staple edibles – even of carbohydrate providers – went far beyond acorns, depending on place, habitat and season. Walnuts and apples were used in central Asia, sweet chestnuts and olives in the Mediterranean, hazelnuts in northern Europe, reed-mace pollen in parts of North America. Like all overarching theories, Logan's Big Idea requires a lot of fanciful Eureka! moments and highly selective evidence, and its end result, ironically, is to underplay early humans' inventiveness and the tree kingdom's edible diversity.

❧

The big oak at the end of our Norfolk garden is a kind of coda, a flourish of contrapuntal woodiness that says decisively: cultivation ends here. Its canopy is twenty-five yards across, a dome of craggy, arching, algal-tinted ribs. Standing in its aqueous shade is like being inside some immense beached cetacean. Tawny owls, flocks of fieldfares, rising moons, the sentence I was mulling over as I wandered up to look at it, can vanish in a trice in its surf of flickering leaves. It was a long time before I could bring myself to do anything as mundane as put a tape measure round its trunk. Just over nine feet; so probably no more than a century old. That brought its stature down a peg, and for the first time in the ten years I'd lived with it I looked at it purely as a structure. I was slightly shocked to realise that my Romantic maze of wood was unambiguously geometric. Working my way round the trunk, I could see that the main side branches (the first, and biggest, grew southwards, ten feet up) were at right angles to each other, and all left the trunk at close to forty-five degrees to the vertical. Some way along each of these slanting ribs, secondary branches sprung out horizontally, at forty-five degrees *below* the supporting branch. And so it went on to the outermost twigs, an alternation of upward and downward forty-five degree divisions. Even the ribs in the leaves were set at this angle to the central spine. On one branch system I counted nine almost identically angled forks between trunk and leaf tip. Of all the oaks in all of Norfolk I seemed to have the one designed by Pythagoras.

If I'd had more of a surveyor's skills and temperament – and been able to see through the tangles of ivy and bramble – I would doubtless have discovered other intimations of order in what I'd always assumed to be an epitome of anarchic growth. I might have found that the numbers of main branches followed the Golden Ratio, with five in the lowest layer, three in the next, two at the top. This is a proportion found throughout nature, in the spiral form of tornadoes and the efflorescent growth of crystals just as much as in the organic world. It has little to do with Darwinian evolution, but seems to be an inherent property of self-organised systems. There are more patterns consequent upon the laws of physics and mechanics. Leonardo da Vinci outlined one such in a formula which he believed described the organisation of all trees: 'all the branches of a tree at every stage of its height when put together are equal in thickness to the thickness of the trunk', a pattern which ensures that there is always sufficient branch capacity to carry the sap from the trunk. At the end of the nineteenth century, the biologist Wilhelm Roux added some riders to this formula: if a central trunk or stem forks into two branches of equal width, they both make the same angle with the original stem, and side branches small enough to make no appreciable deflection in the straightness of the main stem, diverge at angles between seventy and ninety degrees. In the 1920s, the physi-ologist Cecil Murray attempted to explain this by suggesting that the rules governing blood flow also applied to sap. He was studying animal arterial systems at the time, and saw analogies between these and the vascular structure of trees' water channels. The energy required to

drive blood to the point reached by an arterial side branch is minimised
if narrow branches diverge at wide angles and wide ones diverge at
equal or wide angles. Since trees carried water and sap in a similar
fashion, why shouldn't the principle of minimum effort apply here too?
It seemed a logical explanation, and probably holds for some parts of
some trees. But argument by analogy, as so often, proved to be off kilter.
In 2011 the French physicist Christophe Eloy began to suspect there
was an alternative explanation. By using computer models of trees of
different structures and submitting them to virtual gales, he was able to
prove that trees followed Leonardo's formula – accurate enough in itself
– not so much to ensure efficient transport of sap as to give their frames
stability in the wind. Tree forms follow 'the axiom of uniform stress'.
The stresses they experience have to be distributed evenly over their
whole structure, otherwise there are weak spots. The lean of a trunk is
balanced by the counterweight of branch opposite, for instance, which
is supported by thick, muscular tension wood at the joints.

As for the way the branching pattern is repeated, with the fork of
trunk into branch echoed in the splaying of the twigs, the ribbing of the
leaf, even in the radiation of water vessels, it is the phenomenon known
as fractalism. Structural patterns echoed at an ever diminishing scale
are widespread in nature, from river deltas to snowflakes, and appear
again to be self-organising according to mathematical and mechanical
laws. In plants they're additionally encouraged by the economies of
biology. The amount of DNA required to program plant growth so that
it is based around a single structure repeated at different scales, is less
than that needed to make many dissimilar structures.

Yet in the real world these are only ever ideal models, approxima-
tions to a pattern. Living plants are subject to unquantifiable and unpredictable stresses. They're bent in the wind, raddled by fungus, shaded
by their neighbours. Every Platonic pure intention is overthrown by
the realities of life, which is not to achieve perfect form but to survive.
Underlying patterns must help with this otherwise they would have
failed the trials of evolution, but they have done so by being flexible
guidelines, not rigid moulds. And we seem to prefer it that way. A tree
with exactly geometric branching patterns, its branches as fractally

precise as snowflakes and with every leaf identical, would not strike us as a living thing. Being 'frayed and nibbled', Annie Dillard writes in *Pilgrim at Tinker Creek* (1974), is the price of existence. It is one of the qualities that can make us become emotionally attached to old trees. They wear the arboreal equivalent of lines on the face. And some of those lines are added by us.

There are three notable oaks in my home county of Norfolk which show how the interplay between natural growth and human image making shapes real trees, which in turn shape trees of the imagination What's known as Kett's Oak stands by the side of the road between Wymondham and Norwich. It's reputed to be the tree under which, in 1549, the farmer William Kett gathered his army of aggrieved local landworkers before they marched on, and briefly occupied, the city of Norwich in protest against the enclosure of common land. The original oak would have been about 600 years old by now, a giant barrel of a tree with a girth of more than thirty feet. What stands by the B1172 today is a more shrivelled thing, surrounded by a fence (but a low fence, unlike the Fortingall yew) and barely a quarter of that girth. The upper part of the trunk leans sharply away from the vertical, and is propped up by a wooden brace. The stumpy main trunk (there is just eight feet of it below the slant) is partially split and held together with metal bands. I very much doubt that it is the original tree, though it may be a regrowth from the stump after the original trunk collapsed – or was deliberately destroyed (it was a very *political* oak). I rather prefer this regeneration scenario. Leaning slightly away from the rushing traffic of the modern world, far from magisterial, slightly the worse for wear and affectionately looked after by the local authority, the oak is an apt and powerful symbol of the commonplace rebellion it commemorates. (There was another oak associated with Kett on Mousehold Heath, north-east of Norwich, where his 15,000 strong army was encamped. The Oak of Reformation no longer exists, but legend is that it supported a vast tent in which Kett and his helpers planned their action, like Alexander the Great on a major campaign.)

The Poringland Oak is one of the most celebrated paintings by the Norfolk artist John Crome, a member of the Norwich School, whose members prefigured Constable in their reaction against classicism and their desire to paint naturalistic landscapes and ordinary working people. The picture, made in 1818, is dominated by a young, straight and lightly branched oak. It rises beside a pond in which four village youngsters are bathing, or maybe just paddling. They are half clothed and facing away from the viewer, adding to their sense of rural informality. Whether the tree still exists is uncertain. There is an oak beside a pond in Poringland (an expanding village south of Norwich), which may well be Crome's tree grown on for a couple of centuries, and is championed by some locals as a candidate. The land around the pond has been developed, and the tree rises in the garden of a modern, bungalow-sized Free Church with attached café, faint echoes of Crome's populism. When I look at it I'm struck by how different it is from Kett's. It is still as straight trunked as the oak in Crome's picture, and elegant enough to pass for a beech at a distance. Its form is worthy of a landscape garden or arboretum; the perfect model of a tree. Regardless of their authenticity and origins, Kett's and Crome's oaks represent two cultural types: the workaday oak of the commons, and the oak of the pastoral idyll.

The third emblematic Norfolk oak is on the ceiling of Norwich Cathedral cloisters. It is growing from the face, fine featured and long haired, of a Green Man. To be strictly accurate it is not an oak tree, but four oak leaves. But they have the look of individual trees. They're borne on stems as proportionately thick as trunks, and their edges are crimped and finished with gold leaf, as if the whole tree is already touched with the ambivalent splendour of the Fall. This is the oak of mythic creativity.

There are eight more examples of foliate heads in the cloisters, but not all bear oak leaves. One with the face of a gigolo is shrouded in a mask of gilded hawthorn leaves. Another is cheekily diabolic, with unidentifiable leaf forms emerging alongside a tongue from a leering, beetle-browed visage in what is perhaps the most familiar format of

The Poringland Oak, John Crome, *c.* 1818–20.

the Green Man. The foliate head is a diverse and persistent emblem, whose meaning, or meanings, have been argued about for more than 1,000 years. Basically it consists of a human head, wreathed in or shaded by or occasionally composed of leaves, or with leaves growing out of – or perhaps *into* – its ears, nostrils and mouth. The different perspectives of those two prepositional possibilities show the latitude of interpretation to which the Green Man image is open. It may be a symbol of the devil, of death and new life, of the unity of humankind and nature, or just an enduring motif for a long series of cartoons.

The oldest versions date from the cusp between pre- and recorded history. Representations of the Celtic god Cernunnos show his hair formed from leaves. Sixth-century heads on Byzantine capitals sprout acanthus leaves. The first representation at a Christian site appears to be on the tomb of St Abre (now preserved at Poitiers), which dates from the fourth or fifth century. The latest Green Men – all very oaky – are purchasable as plaster facsimiles for interior decor, where they contribute to a Merrie England atmosphere alongside the real-ale bottles and cricket photos. But classic Green Men are most densely concentrated in north European churches from the tenth to the seventeenth centuries, and undoubtedly have a theological status, if not a single meaning.

Interpretations of the Green Man tend to be either stern or celebratory. Kathleen Basford, in her classic *The Green Man* (1978), takes the admonitory side and views foliate heads as predominantly warnings about the temptations of the physical world. She traces this strain in their symbolism to the influential eighth-century theologian Rabanus Maurus, to whom 'leaves represented the sins of the flesh or lustful and wicked men doomed to damnation'. They stood for corrupt words issuing from the mouth, and licentious images entering the eye. By contrast William Anderson's scholarly and pantheist *Green Man* (1990) is inclusive and accepting. He sees the figure as a universal symbol (the subtitle of his book is *An Archetype of our Oneness with the Earth*). He tracks apparent changes in the style of the heads through time and through the physical structures of the Church. The earliest seem the most satanic. The faces are relentlessly fierce. They have open mouths, bared teeth, protruding tongues. During the Renaissance the heads become softer, and begin to develop the believable features of real people. The vegetation surrounds the face more often than protruding from its orifices. They can be found in all kinds of situations inside churches. Anderson suggests that, when positioned by the choir, their leaves represent the issuing of the Word, in song or litany. Over doors through which the living enter and the dead exit they may be memento mori, reminders that 'all flesh is as grass'.

Yet formulaic interpretations don't fit with the exuberant variety of form and ingenious placing of foliate heads. They can be found high up amongst gargoyles and hidden under choir benches. They show the huge influence of the individual carver's imagination and personal sense of humour, or reverence. There are Green Men which are caricatures of village elders, terrifying portents of damnation, clever visual puns. One of the most beautiful in England, a fourteenth-century carving in Sutton Benger church in Wiltshire, features a face with an expression of patient resignation, and emerging from its mouth are sprays of hawthorn in which two thrushes are busy eating berries. In the church in Brome, Suffolk, ancient but renovated in the Victorian period, the mason has carved one distinguished but rather bland foliate face, and next to it an ingenious cluster of oak leaves, in which the gaps between the lobes appear like the eye holes in a carnival mask.

I've seen many Green Men across Europe and think that over the centuries they developed into an all-purpose design feature, a logo endowed with the perennial magnetism of the chimera, and an irresistible eye-worm for stone carvers. Many were doubtless intended to have religious or spiritual significance, but more, I suspect, were created for mischief or ornament, or because they seemed the most apt image to make at a particular spot in a church. A few appear to be used as a kind of vase, in which to root the stone foliage that wreaths about the walls.

The Lady Chapel of Ely Cathedral, forty miles west of Norwich, has one of the most elaborate leafy interiors in England. Many of the images of Mary on the columns were smashed or decapitated during the Reformation, but the Green Men and the symbolically sinful foliage were puzzlingly left intact. There is tremendous cluster of truly diabolical, leering faces (including, uniquely, a foliate fox mask) on its vaulted

The thirteenth-century 'Green Man of Bamberg' in Bamberg Cathedral,
a foliate face unlike any other.

ceiling, and a head on one of the columns which is more village fool than dirty devil. His lolling tongue turns into a stalk which winds across the chapel wall, morphing into leaves, lobes, fruit, tendrils. It is as if the carvers' imaginations worked analogously to the growth of vegetation – 'taking a vine for a walk', to misquote Paul Klee's definition of drawing. A sombre plaque announces that 'this is a place of brokenness reminding us of the brokenness of our world'. On the contrary, the exuberant carving seems a celebration of the unbroken connectivity of the living world. Outside the chapel there are exact leaf and flower carvings everywhere, not just oak, but maple, strawberry, buttercup, hawthorn – all eventually leading to the lines of floral decoration that rise up on the oak fan vaulting of the octagonal lantern.

That there were obvious comparisons to be made between Gothic architecture and the disposition of wood and foliage in trees fascinated the nineteenth-century writer on art and architecture John Ruskin. His views on leaves were ambivalent. He was repelled by the idea of photosynthesis, which invited us to regard leaves as, in his word, 'gasometers' – a view which on the surface seems on a par with his disgust at the notion that the structure of flowers was for the benefit of insects rather than aesthetically minded humans. The arbiter of Victorian taste saw beauty in nature as a benediction from God on those with the eyes to see. Yet the remarkable passages on the development of foliage in Part V ('Leaf Beauty') of *Modern Painters* are devoted to understanding the needs of the growing and leafing plant itself. He notes how the leaves in the cluster – the 'star' – at the end of an oak twig are never equal or symmetrical. 'Nature cannot endure two sides of a leaf to be alike. By encouraging one side more than the other, either by giving it more air or light, or perhaps in a chief degree by the mere fact of the moisture necessarily accumulating on the lower edge when it rains, and the other always drying first, she contrives it so.' Later, in a section called 'The Law of Resilience' he continues:

The ubiquity and versatility of foliate form.
Oak-leaf carved from snow for a contest whose motif was 'feathers'.

[T]he leaves, as we shall see immediately, are the feeders of the plant. Their own orderly habits of succession must not interfere with their main business of finding food. Where the sun and air are, the leaf must go, whether it be out of order or not. So, therefore, in any group, the first consideration with the young leaves is much like that of young bees, how to keep out of each other's way, that every one may at once leave its neighbours as much free-air pasture as possible, and obtain a relative freedom for itself … But every branch has others to meet or cross, sharing with them in various advantage, what shade, or sun, or rain is to be had. Hence every single leaf cluster presents the general aspect of a little family, entirely at unity among themselves, but obliged to get their living by various shifts, concessions, and infringements of the family rules.

Heartened by this burst of arborocentrism in Ruskin, I went back to the oak in my garden and tried to re-envisage it as a compromise between geometric order and the realities of neighbourhood life. Even at a century old it's wearing the badges of experience. I can see that the first big southward branch has grown to balance out the lopping of others on the northward side, where the tree overlooks a worked field. The next branch up lopes eastwards for fifty feet, rippling like a shallow wavelet as it rises and falls over the remains of a hedge, until it is clear of the mother tree's canopy, where it shoots vertically upwards with the clear ambition of becoming a second trunk. Not a single branch is following a straight line. They've been deflected by past ice falls, beetle invasions, and by what seems like a gratuitous and universal urge to wander about. Here and there I can see more abrupt turns and kinks, and through binoculars I can make out fracture marks and scar tissue where branches have been snapped by the wind and had to renew growth in a different direction. I try to follow a single limb, imagining it as a three-dimensional graph, logging the branch's adjustments to over-shading and manoeuvres into spaces of light and still air. For a few feet it becomes a whirlpool of wood, an 'infringement of family rules'. It bends to ride over the branch beneath, which in its turn is trying to evade the presence of this intrusive sun shade. At the point where the two come closest together there are radical attempts at a breakout, with

U-bends, actual shedding of branch ends and sheaves of compensatory twigs. All this is happening in no more than a cubic yard – a cameo of the whole oak family's protean nature.

MYTHS OF CULTIVATION

ONCE AGRICULTURE HAD BEEN DEVELOPED – largely at the expense of
the forest – and became a prelude to the creation of the first cities, it
threw up new plant mysteries. Why did trees, hacked down like enemy
troops in the business of creating fields, so obstinately grow back? How
did a weed metamorphose into an edible vegetable? What made one
plant kill you, another make you well? The period between the beginnings of Neolithic farming and the advances of evidence-based science
in the eighteenth century was rich in myths and fables which attempted
to explain these puzzles. They may not be 'true' in our contemporary
sense, but are fascinating revelations about how the pre-scientific imagination explained the lives and properties of plants, and fitted them into
a view of the cosmos

From the plant's point of view the baptism of cultivation was a mixed
blessing. It meant, usually, a great broadening of the variety of forms in
which a species could exist, as humans selected or cross-bred for traits
they found desirable. Colours, tastes, hardiness, handsomeness could
all be teased out from the rich potentialities of a species' genome. In
the extreme case of the domestic apple some 20,000 distinct varieties
are believed to have been developed from one aboriginal species. But
these pampered variants are often bought at a cost. The development of
copious fruiting may mean the loss of disease resistance. Exuberance
of petal may mean absence of scent, and therefore of pollinators, as in
many rose varieties. It is rare for a portfolio of vegetal qualities regarded

as valuable by humans to coexist with the characteristics a plant needs to survive, unassisted, in the wild. Most of the millions of modern cultivated plants would become extinct within a generation if humans were to vanish from the planet.

9

The Celtic Bush: Hazel

THE TWO VEGETAL STATES of forest and cultivated field still symbolise non-negotiable human conditions: savagery and civilisation. Yet some places, because of peculiarities of soil and history, seem forever suspended in an in-between state, where each brief advance in cultivation is followed by a retreat, returning the land to a time before human presence. One is the range of ragged grassland and hazel thickets known as the Burren in County Clare, Ireland. The video at the information centre in Kilfenora which introduces this shifting landscape to visitors begins: 'Here, on the farthest western edge of the Old World ...' These words feel more than just a geographical orientation. Looking over this expanse of shattered limestone rock on the cusp of the Atlantic, you have the sense that they are also suggesting a timeline, as if this Old World landscape isn't yet finished, or has floated over Galway Bay from somewhere else. Neat tufts of alpine flowers sit amongst a Celtic garrigue, a tangle of blackthorn, burnet rose, creeping madder. There are a few mature trees, but the defining plant is a shrub, the hazel, and its dark smudges fill the hollows in the white rock, stubble the gentler slopes and strain against collapsing walls. The horizon – a series of hunched glaciated ridges, shelved and terraced, in which the low sun picks out ribbons of pink – is never far away. Auden called limestone landscapes regions of 'short distances and definite places'. The Burren (from the Irish *boireann*, a stony place) is just so: compressed, intimate, candid. Hazel is the binder which holds it together.

I've been there half a dozen times, and each time the sense of it being a definite place but from some other time is stronger. My first trip was in early summer in the 1970s. There were four of us, all mad for flowers, and we spent the week in a myopic ecstasy, hopping about the wild rock gardens and revelling in their intoxicating mix of the Celtic and the southern. We foraged mussels and scraped sea-salt crystals from the lichen-blackened limestone that dipped down into the sea. In the spa town of Lisdoonvarna (where our hotelier let us cook the mussels in his kitchen) we lounged under palm trees and watched the visiting nuns' evening *passeggiata*. Mostly we walked about, barely able to drag our eyes from the ground a few inches in front of our feet and the brilliant profusion of plants scattered about with what seemed like complete indifference to ecological protocol. There were Mediterranean orchids next to arctic avens, cobalt blue gentians and common primroses sharing ground in the shade of dwarf hazel copses. We had the bizarre experience of wandering, like Gulliver, through 200-year-old floral forests that came up no higher than our chests. The hazel woods mark time like a metronome, but always seeming, in their multi-cultural jumble of plants and curious miniaturisation, to keep the beat of a landscape which belongs to another time and rhythm, the halcyon days just after the end of the last ice age.

This casual mixing of plants which have no business growing together in a modern climate is the Burren's great conjuring trick. When the glaciers began their retreat from what was to become the British Isles 14,000 years ago, they left behind a landscape of wrecked rock, cracked and strewn about by long winter frosts and summer meltwaters. The ice had swept the rock clean of humus, and of any opportunities for plants which depended on it. So, as the climate continued to warm, a vast array of species adapted to open and infertile habitats edged north. From their redoubts in the ice-free south they crossed the land bridges which still joined Britain to the continent. In the dry alpine summers, on ground in which there was as yet no great competition for space and no inhibiting shade, species we now think of as confined to specialised habitats grew amicably side by side. Bellflowers and bedstraws, gentians and globe flowers came up from central Europe and the Mediterranean.

Rock roses scrambled over not just chalky outcrops but any patch of dry ground. Cornflowers bloomed on the so-far cornless tilth. Sea pinks grew far inland and mountain avens at sea level. A few species, such as the strawberry tree and some scarce heathers, probably migrated along the western European seaboard, which was still an unbroken strip stretching from Portugal to Ireland. Both streams converged in the Burren, blessed by a mild Atlantic climate, and never departed.

As plant remains began to build up a thin layer of soil, and the temperatures rose still further, so the first trees began to arrive about 9000 BC – birch and pine, soon followed by abundant hazel, the Burren's keystone species. The big forest trees, oak, elm and ash, came not long after. By 6000 BC most of the lowlands were covered by a patchy mantle of mixed deciduous woodland, except that on the thin soils of the Burren, the big trees were never very big or very permanent. They were easily cleared when the first farmers arrived around 4000 BC (they came by boat, as the rising sea had broken through the land bridges joining Britain and Ireland to the continent more than 1,000 years earlier). The new pastoralists introduced a herding system which was a reversal of the traditional transhumance in their European homelands. They took the cattle down to the grassy lowlands in summer, and back up to the rocky uplands in winter, where their animals munched up the remnants of the previous summer's vegetation and helped keep the ground open for the following year's panoply of flowers.

Hazel thrives and fruits best in open, well-lit habitats, but survives perfectly well as an undershrub in closed woodland, and in the Burren persisted in the shade of the oaks and elms. Then it received an unexpected boost. Around 3800 BC there was a sudden and mysterious collapse of the elm population across Britain. Early theories about what is called the 'Elm Decline' were mostly based on the idea that early Neolithic farmers were using elm foliage as fodder for their stock (a practice still followed in parts of Europe). This would have reduced the amount of pollen falling to the ground. Pollen grains are good identifying features of the species that produce them and persist for a long time in the oxygen-deficient and preserving mediums of peat and silt. As layers of these earths can be dated by geological features, pollen relics provide the best evidence for identifying and dating the vegetation of a region at a given moment in history. Pollen cores taken from Diss Mere, just a couple of miles from my house in Norfolk, show that just before the Decline elm made up about 7 per cent of the pollen, and hazel about 15 per cent. In layers dating from a few decades later, hazel comprises nearly 50 per cent of the pollen and elm's is vanishingly small. A few hundred years on, elm pollen has begun to return. But the extent and speed of the Decline make it very unlikely that the tiny population of Neolithic farmers were responsible. Oliver Rackham's more persuasive explanation is that it was due to a sudden, early outbreak of elm disease.

Whatever the reason, the same pattern was evident in Ireland, except that in the Burren the big trees never really made it back. The settlers found it easy to keep them down on the thin soils, and the less obtrusive hazel continued its role as the region's defining woody vegetation. And it demonstrated to early settlers – in perhaps a more graphic way than any other species – the vegetal concept of natural regeneration. Trees might have bulky 'bodies' like animals, but they seemed impervious to injury, maybe even to death. Leaves were shed every autumn and returned in the spring. Lost limbs were replaced. A tree upended by the wind might put up a new trunk parallel to the now vertical root plate. The upward growth and branching pattern of trees was maintained regardless of their orientation. Even when whole trees were felled by the laborious blows of stone axes, they would sprout rings of new shoots from growth nodes round the top of the traumatised stump. When these were large enough to be cut the tree would simply grow more. It's a process which has been called 'the constant spring'.

Trees' capacity to regrow after damage – one of their and other plants' major distinctions from most animals – may have evolved as a quid pro quo to compensate for their immobility. They can't run away. They frequently get eaten, or broken by weather, and therefore benefit from having a modular design with no irreplaceable organs. Many can lose up to 90 per cent of their tissue and still sprout again from a fragment of root or twig. This potential began to develop during the long coexistence of trees and big herbivores. The ancestors of the hardwood trees we're familiar with today were evolving under the pressure of browsing mammoth and bison and hippopotamus – serious mashers of woody tissue. If the trees hadn't developed mechanisms for regrowth, they would have become extinct.

There is strong evidence that humans began exploiting natural regeneration – and then deliberately encouraging it – at least as far back as the Neolithic. The poles that make up the ancient trackways across marshy ground in Britain (some nearly 6,000 years old) are too straight and long to have been cut entirely at random. Their regularity, Rackham believes, suggests their growth was being managed with this end in mind, in an early but by no means primitive form of coppicing. The more they're cut, in a cyclical programme that can continue indefinitely, the straighter and more even their regrowth becomes.

This may have been the way that hazel was grown for the wattle huts of the same periods. But it also, unusually, coppices itself. As well as a cluster of thick poles, which grow upwards in the form of multiple trunks, hazel is continuously putting up sheaves of new straight wands from its base. This is how it grows naturally, even when unbrowsed and uncut. Hazel must have been a favourite material for early Neolithic woodworkers. Its thin poles, invaluable for the making of small things, would have been a relief to cut after the heavy work of axing full-sized trees. And its mode of growth was intellectually stimulating. No other European small tree grows in quite the same way. It has a roominess, an internal space full of potential. As it ages, it expands in breadth more than height. The larger outside poles grow gently outwards. The spontaneous new poles springing from the rootstock and points near the foot of existing trunks, shoot vertically upwards between them. Their tight bunching suggests, seems to prompt, the firewood faggots and wattle frames that humans will make of them. They grow together as if they are 'remembering', or attempting to return to, the trunk template that is part of what can be seen as a tree's essential identity.

The character of poles also variegated from shrub to shrub, and was an inspiration to pattern making. One modern hazel worker told me how the colour and texture of the bark can be a guide to the qualities of the internal wood, genetic diversity, as it were, provoking cultural inventiveness. Some hazels have an 'almost metallic sheen and a fine-grained flaky texture like an even scatter of tiny grains of bran'. This type has a strong grain which is easy to split evenly, and is ideal for hurdles and fences. Another has a smooth ground texture to the bark, 'almost like dark olive-green lacquer work', but is brittle and hard to cleave 'like splitting a stick of rock'.

Hazel's response to the ebb and flow of humans over the past 6,000 years, retreating and retrenching as it was burnt, munched and harvested in times of plenty, and released to grow again during famines, is a shrubby text written across the face of the Burren. The groves – sometimes even single straggly bushes – act like beacons in the open rockscape. When rambling over the limestone I'm always drawn to them, and find I'm following a narrative that tells the whole confused history of the place – its wild beginnings and ephemeral spells of plenty, its cyclical returns of simulacra of post-glacial landscapes. The groves, as often as not, are the punctuation marks of social collapse: living cairns, memorials to ruined settlements and collapsed shrines.

There is an ecological as well as historical pattern to the Burren's vegetation. At the foot of the low hills the rock is often tangled with low blackthorn bushes and burnet roses, whose cream-coloured flowers send wafts of warm honey and vanilla into the breeze. Every lip and crack of rock is different – bevelled, honeycombed, rounded, sometimes as sharp as splintered bone. On my first trip I donned a pair of the then fashionable outdoor shoes called Kickers, and kicked my way clean through the soles in less than a week. Rain can dissolve and carve the limestone into extraordinary shapes – razorbacks, rounded gargoyles, Moorish irrigation canals, fantastic zoomorphs – then redeposit the stone as tufa worm casts or facsimile fossils. Hart's-tongue ferns haunt the deep fissures, often growing extravagantly tall in their search for light. There are ash trees too, maybe hundreds of years old, creeping horizontally along crevices; others topiaried by grazing animals into almost perfect hemispheres.

Early purple orchids in a hazel grove in the Burren, County Clare.

Every so often the hills flatten out into plateaux, limestone pavements which were scoured of most of their turf by the glaciers (you can still glimpse scratch marks made by ice-driven pebbles on some of the rocks), and it's on these slabs that the incongruous mixture of the Burren flora is richest. In late May and June they are quilts of bloody cranesbill, rock roses, mountain avens, and as many as twenty-five species and subspecies of orchid. The Burren's orchids are an epitome of the region's geographically inclusive flora. Near Mullaghmore, the dully hued and dully named dense-flowered orchid, *Neotinea maculata* (in Irish called simply *magairlín glas*, 'green orchid' for its narrow spike of drab, dwarf flowers), which has its other heartland on the limestone of the Mediterranean, grows next to a dramatic variety of the early marsh orchid, usually found in Scandinavia and the Swiss Alps. Watching two smart botanists debate the identity of this prodigy, *Dactylorhiza incarnata* subsp. *cruenta*, gave me a new view of the usually austere science of taxonomy, with its reductionist counting of pistils and esti- mations of hairiness. Just at that moment it seemed to be providing a portal into the beauty of small detail. Bob and John were crouched over the plant with the attentiveness of stoats contesting a rabbit, but in the kind of amicable conversation two music scholars might have discussing the interpretation of an early manuscript. *Cruenta* (it has no English or Irish name) is a striking plant with a tower of lush purple flowers, and blade-like leaves streaked and stippled with extraordinary dark, almost creosote-coloured blotches and curves. But there are other marsh orchids with spotted leaves, and the recumbent pair were deter- mined to nail its identity. They had their high-end orchid guides open on the ground, and with their lenses worked their way through the defining parameters of foliage and flower. Leaf shape? Elliptical to lanceolate, usually hooded at the tip. Markings? Dot- and loop-marked, usually shallowly trilobed. Bracts? Relatively short, tinged purple and *spotted*. It was a true *cruenta*, and there was much rejoicing.

Why should its identity matter? The attachment of a precise name is irrelevant to the lives of plants themselves, which are continuously evolving new varieties (and therefore new labels) in response to changes in circumstance. Orchids especially are a promiscuous tribe whose members are always throwing up novel forms and hybrids. But I'm comforted by the act of naming. It's a kind of befriending, a recognition, however temporary, of individuality and provenance. *Cruenta* grows in only one or two other sites in Britain, in the remote Scottish Highlands. To have seen it in the Burren, this archive of particularity, made me feel oddly privileged. Its identity gave those cryptic markings the aura of an imprimatur, a stamped confirmation of the mysterious territorial loyalties of plants.

You could see the expansive and evocative nomenclature of plants as another expression of the efflorescence of the imagination, our human response to the vegetal world's profusion of spots and lobes. Hazel, *Corylus avellana* in Latin, is *coll* in Irish, or *airig fedo*, 'a noble of the wood'. Hazels in sacred sites or with special powers are called *bile*. In Gaelic poetic theory, *caill crínmón* are the 'hazels of scientific composition' referring to the hard nut, from which a new composition is broken open or unshelled. The yews which grow, often dwarfed and hunchbacked on the limestone pavements, are *iubhar*, a name which links them with the clan of European *iws* and *yws*.

The Burren's physical landscape is sonorous too, naming itself in a language of dry scrapings and abrupt consonants. Desiccated lichens and rose leaves crisped by the sun crunch under your feet. Loose stones sway and clatter – a bony, ringing, almost tonic note. On one June walk I heard a few words floating from the tannoy of a passing tourist coach: 'Four hundred years ago the Burren ...' I never caught the rest. The past here seems vaporous, pregnant, dust of the same order as the seed heads of gentians and orchids. For all its ancient echoes the land is overpoweringly present, fresh minted, and in the glaring sun reflected off the white rock has the look of an emerging image on a photographic print. I'm tugged this way and that by mirages and things that shimmer at the edges of my vision. A pink burnet rose. A pool of distant yellow which turns out to be a prostrate golden-leaved holly. A *trompe-l'oeil* model of a Bronze Age wedge tomb, six inches high, which some ironic rambler has propped up on a slab. I need to make an effort to stop and perch, and even then the place still seems to be on the move. Flakes of mica-thin rock, limestone millefeuille, flutter down the terraces. The breeze ruffles thin puddles of water on the slabs – 'tadpole runnels' – that will, slowly but inexorably, dissolve their way through the limestone to make new crevices. These will become, in effect, the templates for shady, miniaturised gorges, subterranean woods, and in them a new generation of woodland ferns and flowers – and hazels – will take root.

'Four hundred years ago,' the tourist coach guide might have continued, 'the Burren was going through one of its more prosperous times.' The cattle would have been thriving, the hazels subdued, cowering in the grikes and shooting horizontally out of rock faces inaccessible to animals. Two centuries later the region was struck by the great Irish famine, and the hazels would have re-emerged. Over the centuries homestead and hazel have pushed and pulled, gained and lost ascendancy, their fortunes depending on weather, politics and the respective obstinacies of shrub and shepherd. Near Poulbaun I find the remains of an undateable farm holding. Slivers of stone, like hands and crosses, have been jammed upright in the crags above the crumbling relics of the field walls, now barely distinguishable from the natural tiers in the limestone. The Burren's great biographer and mapper, Tim Robinson, describes them as 'memorials to the tedium of the herdsman's life'. A few miles on, just south of Kilnaboy, the relics of a triple cliff fort, Cathair Chomain, circa AD 1000, look scarcely more durable. The hazel woods are creeping up the hill towards it and the stone fortifications slowly collapsing into the scree. There are four millennia of abandoned dwellings and monuments here – ring forts, dolmens, holy wells, sixty-six megalithic tombs, 500 ring forts and countless hovels and goat huts. Just below Cathair Chomain is the collapsing shell of a lurid farm bungalow, built in pink-and-yellow bricks circa 1958, and a few hundred yards further on a completely houseless garden, where the hazel is reclaiming a pergola in which a shrine to the Virgin Mary nestles. Occupation has always been casual and ephemeral here. People settle in numbers, endure trial by rock and famine and then just walk away, bequeathing their hard-won estates to the hazel.

Near another ring fort at Cathair Mohr I shoulder my way into a grove that looks primeval, as if it has never been taken into a farm, or nibbled to the ground by sheep. It may be an illusion: hazel can grow tall and dense within a century. But it is awesome inside, hushed and humid, and seems too physically hostile a location ever to have been anything other than a wood. The hazels are bunched close together and some of them are nearly two yards wide at the base. The floor is strewn with immense boulders and fallen branches, and swaddled with a blanket of moss, eight inches deep in places, which forms thick ruffs up the hazel trunks as if this were a tropical cloud forest. I creep about as gingerly as I can, hoping to avoid the treacherous gaps between the boulders without causing too much damage to the plants. But the moss shows deep accusing pits where my feet have been. For a moment I'm seized by the fancy that I am the first person ever to have trodden in this corner of the wood. I find a glade near the centre and feel as if I'm in an ornamental bower, not a wildwood. It's that aura of the cave again, the sheltered cell. The primroses and bluebells and early purple orchids aren't growing in masses, as they do in woods where the ground is less tenanted, but in delicate ones and twos wherever the moss is compact enough. And in the dappled shade at the edge of the grove I find my very first specimen of the Irish spotted orchid, a pure white spike splendidly named *Dactylorhiza fuchsii* var. *okellyi*, after the Burren nurseryman from Ballyvaughan who first discovered it.

Some of these hazel groves are truly ancient. The most celebrated and most elfin is, in the Irish botanist Charles Nelson's words,

> the greenwood that hangs like a curtain from the 'frowning' cliff called Cinn Aille, draping the scree with hazels, helleborines, ferns and mosses. Scattered among the nut trees are sallys [willows], ash, and rowan. The tiny oratory of St Colman Mac Duach, once a snug retreat but long-since abandoned, is also silent. Its dressed stones are following that ineluctable trail of all the Burren's limestone back to their source, dissolving in the rain and tumbling to the ground to become part of the natural rock-garden.

When I was an earnest young forager, I realised that the most successful way to discover hazelnuts was to be in just such a position – inside a grove, or even an individual bush – looking out, so that the nuts were highlighted or silhouetted against patches of sky. Hazelnuts figure richly in Irish folklore 'always [as] an emblem of concentrated wisdom, something sweet, compact and sustaining, enclosed in a small hard shell: in a nutshell, so to speak'. The story whose compact symmetry is most satisfying concerns a sacred well surrounded by nine hazel trees, representing wisdom, inspiration and poetry. The leaves, blossoms and nuts of the trees all break out at the same moment, and fall into the holy water, raising a purple spray. Five salmon living in the well would eat the nuts and for each nut a red spot would appear on their skin. Any person who ate one of those salmon would understand wisdom and poetry. The nuts would also raise bubbles of inspiration in the streams flowing from the well, which would be imbibed by artists and thinkers of all kinds. It is a complicated myth, involving a tree of knowledge and several kinds of metamorphosis, and it oddly echoes the real story of the fruit–river–fish life cycle that sustains some Amazonian forest species. Myths about plants, however superficially mystical, often have buried within them essential ecological truths. The metaphor of the nut as a symbol of condensed wisdom from which some new form will 'hatch' has stayed with us. Yet the image of the hazel bush as a model for regeneration – visibly true and mythically powerful – has been largely passed by or forgotten. The fading of our cultural and workaday intimacy with trees has allowed us to anthropomorphise them in a directly corporeal way, seeing the trunk, the body of the tree, as analogous to a human body, and therefore killed when severed from its earthly roots. This is metaphor badly used. We lament the felled figure and fail to see the aspiring shoots – as satin sheened as salmon – rising from its foundations.

10

The Vegetable Lamb: Cotton

IN THE MIDDLE AGES it was not thought in the least improbable that plants were capable of changing into other beings – maybe not humans, as in classical mythology, but certainly into other kinds of organic growth. How else could new forms emerge, or be explained? In the latter half of the fourteenth century a story began appearing in western Europe about a creature from Tartary (then comprising most of central and north-eastern Asia) which was half plant, half animal. A chimera, like the Green Man. In the notorious *Travels of Sir John Mandeville* (circa 1357), a book which was almost certainly written not by the St Albans-born knight, but by a Flemish monk who himself may have been no more than a collector and plagiarist of others' adventurous tales, the more fantastical and fanciful the better. (It was typical of the book, and the times, that Mandeville admitted he hadn't actually been to the Garden of Eden himself but was certain it still existed.) Whatever its origins, the *Travels* included an account of this fabulous organism, which Mandeville claimed to have seen in 'the Lond of Cathaye, towards the high Ynde':

> There growth there a maner of Fruyt, as though it weren Gowdres: and when thei ben rype men kutten hem ato, and men fynden with inne a lytylle Best, in Flesche, in Bon and Blode, as though it were a little Lomb, with outen Wolle. And Men eten both the Frut and the Best; and that is a great Marveylle. Of that Frute I have eaten; alle thoughe it were wondirfulle, but that I knowe well, that God is marvellous in his Werkes.

It sounded like a parody of an Old Testament parable: the Lamb of God as an edible vegetable.

More than two centuries later, the usually more reliable French botanist Claude Duret devoted a whole chapter of his *Histoire Admirable des Plantes* (1605) to 'The Boramets of Scythia, or Tartary, true Zoophytes or plant-animals; that is to say, plants living and sensitive like animals.' He made no claim to have seen the vegetable lamb, but confirmed Mandeville's story, and filled in more details from a fifth-century Hebrew text, part of the collection that contained the first written accounts of the Genesis creation myths. In Hebrew the creature was called *adnei hasadeh* (literally, 'lord of the field'). It was shaped like a lamb, and had a stem growing from its navel which was rooted in the ground. According to the length of this trunk or stem, the lamb was able to devour 'all the herbage which it was able to reach within the circle of its tether'. Then it expired.

Explorers and writers began discovering – and elaborating – more details of the creature's extraordinary existence. It had four limbs and cloven hooves. Its skin was soft and covered with wool. It had no real horns, but the long hairs of its head were intertwined like vertical pigtails. The wolf was the only predator of the Borametz (a Tartar word signifying lamb), but humans seemed well acquainted with its gourmet qualities, which, in keeping with its chimerical nature, crossed biological divides. Its blood was as sweet as honey. Its flesh tasted of crab, or possibly crayfish. Baron von Herbenstein, German ambassador to Russia, passed on (the accounts are invariably second or third hand) what he regarded as a story that vindicated the Borametz's woolliness. 'I was told by Guillaume Postel, a man of much learning, that a person named Michel' had assured him that he had seen near Samarkand 'the very soft and delicate wool of a certain plant used by the Mussulmans as padding for the small caps which they wear on their heads, and also as a protection for their chests'. By the end of the sixteenth century the story had entered Christian iconography. The French writer Guillaume de Saluste, in his 1578 poem *La Semaine*, imagines the plant as one of those creatures that excited the attention of Adam as he wandered through the Garden naming things. He makes a kind of riddle out of its paradoxical properties.

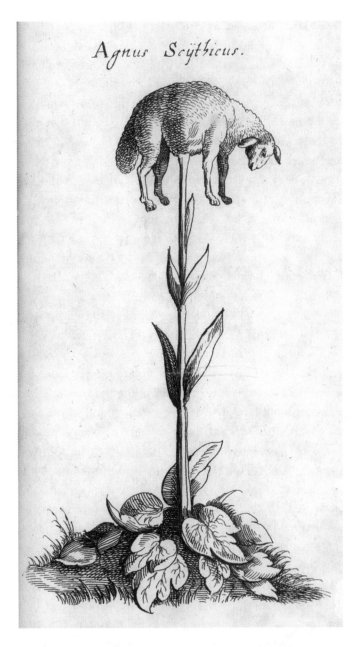

Agnus Scÿthicus.

A seventeenth-century illustration of the vegetable lamb, by Matthäu's Merian the Younger. A confused picture of a confusing myth, as this lamb could not reach the pasture beneath.

The nimble plant can turn it to and fro,
The numbed beast can neither stir nor go,
The plant is leafless, branchless, void of fruit,
The beast is lustless, sexless, fireless, mute:
The plant with plants his hungry paunch doth feed,
Th'admired beast is sown a slender seed.

In Britain, the flyleaf of the early editions of John Parkinson's great seventeenth-century horticultural manual, *Paradisi in Sole Paradisus Terrestris* (the title is a clever bilingual pun: it means Park-in-sun's park on earth), is adorned with an illustration of 'Adam and Eve admiring the plants in the Garden of Eden'. There, amongst a luxuriant collection of apples, palms, lilies, cyclamens, tulips and a ravishing long-haired Eve, is an unmistakable vegetable lamb on its stalk, still with a lot to eat.

With the dawning of the Enlightenment, more sceptical eyes were turned on this extravagant legend. A few shrivelled relics, reputedly onetime Borametz, turned up in European museums, but none of their donors seemed to have set eyes on the living source. It was invariably a story passed on by a colleague who'd read it in some armchair explorer's memoirs. The suspicion grew that it was an ordinary but previously undescribed plant or artefact, badly reported, and hyped up by second-hand rumours and Tartarian whispers. The first scientist to challenge the myth was Dr Engelbrecht Kaempfer, a surgeon for the Dutch East India Company in the late seventeenth century. His searches for the 'zoophyte feeding on grass' were fruitless, but he discovered a breed of sheep local to the area round the Caspian Sea which had exceptionally soft skin and fine wool. There was a local tradition of killing pregnant ewes to harvest the even finer fleeces of the embryonic lambs. Their skin was as thin as vellum. It shrunk on drying, and 'assume[d] a form which might lead the ignorant and credulous to believe that it was a woolly gourd'. Kaempfer believed that it was these desiccated skins which had found their way into European collections to be presented as examples of the zoophyte's fleece.

A more plausible explanation appeared when the arch collector Sir Hans Sloane exhibited another withered object to the assembled Fellows of the Royal Society in 1698. For the next two centuries this artefact was regarded as the source of the entire vegetable lamb mythology. It was part of the root (rhizome) of a tree fern from China, about a foot long and as thick as a man's wrist, and had been crudely whittled, so that it appeared to have a sheep-like head and four legs cut from the stalks. It was covered with yellowish down, a quarter of an inch thick in places. A fabulous plant made out of a fabricated plant. During the eighteenth century, more specimens of these carved rootstocks were seen in Asia, all belonging to the tree-fern genus *Dicksonia*. They were made as toys and curiosities, and doubtless passed off to gullible tourists as something more mysteriously oriental.

And there the matter rested to everyone's satisfaction for another hundred years – bewilderingly, since it stretched credulity that a crude gewgaw could be the source of an elaborate and widespread fable more than 1,000 years old. But in the 1880s, Henry Lee, Fellow of the Linnaean Society and myth-busting author of *Sea Monsters Unmasked* and *The Octopus, or the Devil-fish of Fiction and Fact*, believed he had found the true solution, and one which had a particular resonance with Britain's current imperial ambitions. He poured derision on the ferny figurines. 'I have to express my very decided opinion that they and the "lambs" (?) made from them had no more to do with the origins of the fable of the "*Barometz*" [sic] than the artificial mermaids so cleverly made by the Japanese had to do with the origins of the belief in fish-tailed human beings and divinities.' Writing from the redoubts of the Savage Club in London, a bolthole for travellers and intellectual eccentrics, he offered a radically different interpretation, still rooted in the concept of mistaken identity but more plausible than any previous attempt. Reviewing all available evidence, he concluded that we should now be able to recognise with confidence the vegetable lamb's true 'forms and features under the various disguises it was made to assume by the wonder-mongers of the Middle Ages', and appreciate that it was 'a plant of far higher importance to mankind than the paltry toy animals made by the Chinese from the root of a fern'. The original Borametz, he declared triumphantly, was none other than the pod-bearing cotton plant. It was a satisfying and patriotic verdict from a man who, in his closing chapter, makes an

The fleece-like seedheads – bolls – of the cotton plant.

impassioned plea for Britain's retention of India and of the huge wealth represented by its cotton plantations.

What is odd is that this Byzantine myth should have taken root in Europe at all. Cotton was a familiar plant. It had been imported by southern Europe since classical times, and was a common commodity across the whole continent by the fourteenth century. By the late 1500s fabrics were being made in Britain from raw cotton by Flemish spinners and weavers. The herbalist John Gerard, writing in 1597, was nonchalant about the plant: 'To speake of the commodities of the wooll of this plant were superfluous, common experience and the dayly use and benefit we receive by it shew them.' He tried to raise it in his garden in Holborn. It 'did grow verie frankly, but perished before it came to perfection, by reason of the cold frosts that overtooke it in the time of flouring'.

Hardly a mysterious alien, then, either as raw pod or as finished fabric. Even if the parent plant was unfamiliar, it's implausible that anyone seeing it for the first time – an unassuming bush bearing seeds enclosed in fig-sized bolls of fluff – could mistake it for some phantasmagorical plant–animal hybrid. But the origins of the fable, or perhaps the psychology behind it, resonate with the true history of the cotton plant, which is almost as extravagant as the fictional myth of the vegetable lamb.

There are four members of *Gossypium* genus (part of the mallow family) which are known as cottons, two in the Old World and two in the New. They have been cultivated and hybridised for so long that, in the case of the Old World species, *G. arboreum* and *G. herbaceum*, no truly convincingly wild ancestor has ever been found. The cultivars may have originated in Africa even before the Stone Age, from one of the wild *Gossypium* species still found there, and their wool used in ways unrelated to cloth making, perhaps as a ceremonial adornment, or a staunch for wounds. At some stage these early cultivars were taken to Asia, and evidence of the first expert weaving is from about 1000 BC, in the Indus Valley. They spread slowly west, and in the Book of Esther,

King Solomon's palace walls are described as being covered with cotton fabric. In the fifth century Herodotus, one of the more reliable classical commentators, reported, 'The Indians have a wild-growing tree which instead of fruit produces a species of wool similar to that of sheep, but of finer and better quality.' You can see the germs of a vegetable-lamb myth in his description, but no sense that Herodotus thought cotton anything other than an ordinary plant with unusual accoutrements. Five hundred years later Pliny the Elder finds it has migrated further west, and recognises the pods as, botanically speaking, a seed-containing fruit: 'The upper part of Egypt, facing Arabia, produces a shrub which is called gossipion ... The fruit of this shrub resembles a bearded nut, which contains a soft wad of fine fibres which can be spun like wool and which is unsurpassed by any other substance in respect of whiteness and delicacy.' Cotton continued to follow a familiar track west, through Roman imports of fabric from India to the introduction of plants and cultivation techniques by the Moors to Spain and Sicily, around the start of the tenth century.

Then the story takes an unusual twist. The early Spanish colonisers of the Americas took cotton with them, only to find that the indigenous people were already making sophisticated garments from it. When Hernando Cortés arrived in Mexico in 1519, he was presented with a gold-encrusted ceremonial cotton robe by the Yucatan Indians, just before his soldiers set about slaughtering them. Recent archaeological evidence suggests that two different New World species had been in use there for thousands of years. Seeds of *G. barbadense* discovered in excavations in Chile and Peru, and the remains of yarns and fishing nets made from its silky seed plumes, have been dated back to 4000–3000 BC, and *G. hirsutum* seeds found in Mexico to 3500 BC. It looks as if cotton had been a cultivated crop in Central and South America for at least 2,000 years longer than in Asia. Then, as genetic science advanced in the 1960s, a bewildering discovery was made. The American cultivated cottons had originated in some ancient cross between the cultivars of the Old World and the wild species of the New. How could this have happened, long before European settlers took their species across the Atlantic?

This is where fact starts to wander into speculation, or what might be called scientific fabulation. One group of botanists have theorised that the seeds of Asiatic cottons found their way across the Pacific millions of years ago, carried by birds or ocean currents. They hybridised with the New World species and then became extinct themselves, leaving their vigorous mongrel offspring to continue the line. Another group, the 'diffusionists', argue that cultivated Old World cottons were brought across the sea by the peoples from South East Asia who first colonised the western coasts of the Americas. Yet another group, the 'inventionists', maintain stoutly that hypothetical foreign introduction is improbable and that native American cotton species simply underwent a spontaneous chromosomal mutation. This botanical debate reflects the ongoing and fiercely partisan argument about the origins of Native Americans. When did they arrive, and how? Across land from the northern tundra, or by boat from the southern Pacific? The American botanist and expert on the history of economic plants Edgar Anderson once remarked that European ethnographers are more likely to favour the transoceanic diffusion theory, and their American counterparts to patriotically 'explain our indigenous high culture as a flowering of the aboriginal American intellect. A generation ago, when leading American writers were more uniformly anti-diffusionist than at present, a witty English anthropologist referred to this theory as "the Monroe Doctrine of American Anthropology".'

None of these modern discoveries cast much light on the substance of the vegetable-lamb myth, originating as it did thousands of years earlier. But they may say something about its origins and enduring shape. It's clear that story making is compulsive in all societies and at all times, and that the idea of hybridisation, be it between cultures or organisms, is seductively ambivalent. Cotton is an eccentric plant, tailor made to be a provocative subject on both counts. (The fibrous fluff that embeds the seeds has ethereal qualities of its own, as a kind of vegetable mist. In the heyday of spiritualism in the late nineteenth century softly lit plumes and clouds of cotton wool were routine stage props at séances, intended to dupe the faithful into believing they had glimpsed ectoplasm or spirit 'auras'.)

Myths and fables don't spring into full narratives directly from the collective unconscious, despite mystical beliefs to the contrary. Even if Claude Lévi-Strauss is right when he suggests that a propensity for myth making is hard-wired into the modern human brain, just as the basic structure of language seems to be, at some point myths have to be articulated, turned into stories by individuals with a gift for narrative. Cotton would have provided promising material for fable spinners as far back as the Neolithic. Ordinary village storytellers also doubtless wove tales around its curious anatomy, to serve as parables or entertainment or for simply making mischief with strangers. The stories would have been progressively ornamented as they passed from teller to listener, yet always maintaining two essential parable-shaped motifs – the idea of the plant–animal chimera, and its collapse when its limited food was grazed out. It would have been various versions of these tales, not some mass delusion, which early European travellers picked up. And their long survival suggests that the myth's core struck some sympathetic echo in the societies in which it was first told as well as in the wider world.

Jim Crace's historically telepathic novel about the final days of the Neolithic era, *The Gift of Stones*, has as its central character a storyteller who cannot contribute to his small community's trade of knapping and finishing flints because he has lost an arm. So he becomes a professional spinner of yarns, a polisher of tales. He enchants his fellow villagers with outrageous stories, mostly based on romantic and fantastical riffs on their humdrum, stony lives. He explains what makes for a good story, which if successful enough will take root as a fable. 'Why tell the truth when lies are more amusing, when lies can make the listener shake her head and laugh – and cough – and roll her eyes? People are like stones. You strike them right, they open up like shells.'

But enduring myths are not exactly lies. The ex-Bishop of Edinburgh Richard Holloway suggests that 'the question we should ask of a myth is not whether it is true or false, whether it did or did not happen; but whether it is living or dead, whether it still carries existential meaning for us in our time'. Chimeras figure so prominently in mythologies that they may carry universal meanings, be the kind of 'striking' stories that 'open [us] up like shells'. Hybrid creatures like the centaur, phoenix, griffin and sphinx occur in myths throughout the world. Often they seem to have some kind of real-world analogue. The double helix of DNA was prefigured in an image on a Levantine libation vase from 2000 BC. It shows two great snakes, also entwined in a double helix, and symbolising the original generation of life. The great biological essayist Lewis Thomas describes a Peruvian deity painted on a clay pot from about 300 BC, believed to be a charm for protecting crops. Its hair, too, is made of entwined snakes. Plants of several kinds are growing out of its body and some sort of vegetable from its mouth, Green Man style. Thomas points out that it is an imaginary version of some real animals, species of *Pantorhytes* weevils recently discovered in the mountains of New Guinea. They live symbiotically with dozens of plants – lichens and mosses especially – growing in the cracks of their inch-long carapaces, miniature forests which contain a whole ecosystem of mites, nematodes and bacteria browsing on the foliage. Thomas has nicknamed them *symbiophilus*. We are a kind of chimera ourselves, he wrote, 'shared, rented, occupied'. Hundreds of species of bacteria inhabit our skin and gut (where they're pleasingly called 'flora') and are essential for the efficient working of the digestive and immune systems.

The myths and symbols which echo and maybe intuit the workings of these real-world communities seem to celebrate the tendency for living things to join up, form partnerships, live inside each other for mutual benefit. Human cells are formed from collections of other organisms, living together but independently inside these enclosed bubbles. Our myths of chimera may be born out of the optimism of this idea, or an ancient understanding of it. The story of the vegetable lamb is resonant and densely layered. It's a parable about natural economy and the inseparability of plants and animals. In its evolution it toyed with Judaeo-Christian icons and ideas: the Tree, the Fruit, the Lamb; 'all flesh is as grass'. The vegetable lamb was a beast which, like a plant, had an umbilical connection to the earth. A plant which, like a beast, lived by grazing and died when it had exhausted its food supply.

11

Staff of Life: Maize

THE MOST TRAUMATIC CHANGE in early cultures was from hunter-gathering to agriculture, a shift which involved a fundamental transformation of the landscape from an open prospect dominated by wild trees to enclosures of cultivated crops. The development of maize – with rice and wheat, one of the world's three main staple grasses – was the most remarkable process of informal plant breeding in the history of cultivation, involving mysterious transformations of the plant itself, and empathy and ingenuity on the part of the Indian farmers who lived with it. The myths of corn's origin understandably contain equal measures of nostalgia for the bountiful forest and gratitude for the equally bounteous new crop. In Mayan culture in central America, the origin myths were chiefly centred on Thipaak, a heroic figure who saved his people from having to sacrifice their children, and who was the first to cultivate corn. He stored it in a huge pillar supporting the sky. T'ithach is the name of a 9,000-foot peak in the Huasteca region. Long ago a bird drew attention to maize kernels being carried out of the pillar by leaf-cutter ants. The local rain god struck the pillar with a blast of lightning, opening it up and making maize available to the people. This is a myth of gratitude for maize, tinged with tropes of the hunter-gatherer's life. So too are the group of stories collected by Claude Lévi-Strauss in the Brazilian region of Mato Grosso, thousands of miles to the south. There the beginnings of cultivation are detailed exactly, and there is a sense that agricultural toil is a punishment for a mistake made in an arboreal paradise, echoing the myths of Genesis – 'in the sweat of thy face shalt thou eat bread'. 'These myths,' writes Lévi-Strauss,

> refer to a time when men knew nothing of agriculture and fed on leaves, tree fungi and rotten wood before the existence of maize was revealed to them by a celestial woman, who had taken the form of an opossum. The maize was like a tree in appearance, and grew wild in the forest. But men made the mistake of felling the tree and they then had to share out the seeds, clear the ground for cultivation and sow maize because the dead tree was not sufficient for their needs. This gave rise, on the one hand, to the different varieties of cultivated species (in the beginning they had all been together on one tree) and, on the other hand, to the differences between people, languages and customs.

The myth, in a rather anthropological way, links the evolution of maize varieties with the elaboration of societies and cultures.

There is no single scientific theory of the origins of cultivated maize, or corn, to set alongside these myths, but the increasingly accepted story echoes the Mato Grosso narrative. The wild ancestor, the ur-maize, is now believed to be an annual grass called teosinte, native to dry and waste places in Mesoamerica. Growing to several feet in height, it has branch-like side stalks which carry the narrow ears; these are open, and scatter their seed as soon as it is ripe. The seeds are consequently hard to gather, and vulnerable to grain-eating animals. After decades of study, during which the archaeological evidence seemed to point to at least three separate sites of domestication, it now looks as if the shift from teosinte *Zea mays* ssp. *parviglumis* to maize *Z.m.* ssp. *mays* occurred in the Rio Balsas region of western Mexico about 9–10,000 years ago. The oldest physical remains of a maize cob, found in a bat cave in Oaxaca, dates from about 5400 BC. But pollen deposits and studies of the rate of genetic variation point to a much earlier date.

A likely scenario is this. Central Americans may have first used teosinte for its high sugar content, chewing its stalks – which resemble sugarcane – as modern western Mexicans continue to do. During the Neolithic period, with a developing understanding of the concept of culti-vation, they transferred selected teosinte plants to their rudimentary

A string of dried corn cobs in a house in Tarma province, Peru, showing their variety of shapes and colours and suggestions of symbolic importance.

'forest gardens' and tended them by watering and weeding. They doubtless made some use of the seeds as food, but they were small and difficult to harvest in quantity. But teosinte is a highly variable plant, and mutants would have appeared in which, for example, there were four rows of kernels instead of one, and others which had the beginnings of the protective sheath of modified leaves so prominent on modern corn 'cobs', which meant the seeds weren't scattered on ripening. This also meant the new strain was sterile, unless the kernels were deliberately planted out. This is what must have happened. The Neolithic farmers would have noticed these desirable variations, and selected their seed for next year's planting. In the process they would have made spontaneous cross-breeding between selected varieties more likely. The first cultivated maizes derived from this over a comparatively short period of time, but the crop was destined to be always dependent on the interventions of cultivation for its survival.

These early maize ears weren't much like modern corn cobs. They were stubby and thin. The kernels were hard and gritty. But *Zea mays* is prone to a good deal of diversity, and highly promiscuous. In cultivation, sports (spontaneously different forms) often crop up in which the kernels are coloured, or fatter, or lie in multiple rows, or possess high levels of sweetness. Some of these would have been preferentially selected for cultivation. On the open ground at the edge of settlements yet more varieties would have appeared because of natural cross-pollination, or back-hybridisation with wild teosinte. So as maize began to radiate geographically across central America, and then into the northern and southern reaches of the continent, it also radiated genetically, as different forms were favoured and propagated to meet different tastes and cultural needs. Its spread through human and genetic space has been one of the longest and most complex exercises in amateur (and often unintended) plant domestication, yet one which would have been impossible without maize's own genetic inventiveness.

Eventually, maize reached Amazonia and the foothills of the Andes. And somewhere here the first sweetcorn appeared. This variety was nothing like its modern descendants, the fat cobs we eat in the unripened, 'green' state. It had ears with small, wizened kernels, which contained unusually high levels of sugar compared to their starch content. They turn unappetisingly gummy on being cooked. But they were highly prized in Peru and Bolivia for fermenting into maize beer, or *chicha*, important in religious ceremonies and as a recreational drink. Modern Peruvian sweetcorn destined for *chicha* is a bizarre crop when compared to conventional corn. The cobs are almost spherical, as big as an orange, with many rows of irregular kernels which each taper to a point. They vary in colour from pale lemon yellow to deep Chinese red. These nuances would have begun to appear in the early fields and forest gardens of South American Indians, before making their way north again as a new contribution to the maize gene bank of meso-America.

Forest gardening can be thought of as a conceptual halfway house between hunter-gathering and full agriculture, although the gardens are perfectly sustainable in their own right. They still exist in pre-industrial societies across the globe, and have a basic common structure wherever they are. A small area of forest is felled, and some of the fallen timber and foliage is allowed to remain in situ, as are any useful or ritually important plants like brazil nut and coca. The remainder of the timber is burnt and the ashes used, highly selectively, as a fertiliser. Inside this matrix of rotting wood and leafy shrubs and carbonised branches, staple crops such as manioc and yam are planted. In Amazonia maize is always planted in patches towards the edge of the garden, because its life cycle is different from those of the indigenous crops. There may also be a scattering of leafy greens, pineapples and plants whose crushed leaves are put in rivers to anaesthetise fish. In Colombia, the Tukano Indians don't regard their gardens either as 'clearings' of the forest, or as oases of 'cultivation' in an otherwise hostile environment. They are what anthropologist Gerardo Reichel-Dolmatoff calls a 'safe-hold', a temporary second home where important plants can be grown *thanks* to the forest. The Tukano's understanding of the nature of fertility in the rainforest echoes that of ecologists. The organic debris brought down from the canopy is a crucial ingredient of cultivation. It 'contains nutrients ("energies") not present in the soil, but which come "from above", and which provide

A European's view of Indian 'forest gardens'. John White's bird's-eye 'View of the Indian Village of Secoton' *c.* 1570.

the cultivated species with an articulation with the forest environment'. The notion that, in rainforest ecosystems, 'fertility' exists in and is circulated through the vegetation, not the soil, is still not understood by many modern agronomists.

The gardens are selectively weeded, with any potentially useful seedlings retained, so they would have been ideal places where new sports of maize, sprouting in the open soil, could be singled out for propagation. The likely precursors of forest gardens were areas of disturbed ground near the temporary settlements of nomadic hunter-gatherers, especially fisherfolk and early pastoralists, where edible plant waste and animal dung would have accumulated, conditions which were highly conducive to the generation of new varieties. The distinguished US botanist Edgar Anderson developed a 'refuse heap' theory of the origins of agriculture based on these midden sites. Prehistoric rubbish tips were places where edible plants gathered from different environments could – via their discarded seed remains – grow together in random proximities that would never occur in the wild. With luck they might produce spontaneous hybrids, which would have a better chance of survival in the disturbed open ground than amongst the mature vegetation of forest or grassland. Maybe the small interactive dramas of generation on these heaps – spit out seed, dump dung, churn soil, spot the stranger – inspired the *idea* of cultivation.

During the mid twentieth century Edgar Anderson tracked the way different corns had migrated across America. Moving north along the Andes, he found a range of different drinks made from roasted sweetcorn meal, some fermented, some not. In western Mexico sweetcorn, *maíz dulce*, was still being used as a source of sugar. He found a popular local savoury called *ponteduro*, a kind of Neolithic cracker made from toasted corn kernels, squash seeds and peanuts. In the same region green corn – 'corn on the cob' – was a quite different variety with long, narrow ears, and big blue or red-purple kernels.

Colour seems to have been a significant factor in guiding selection. Individual kernels (normally white to yellow) often appear in various shades of orange and red. Anderson frequently found ears whose kernels were neither solid red nor white, but freckled with streaks and patches of alternating colour. In Mexico these are known as *sangre de Cristo*, literally 'blood of Christ'. He noticed that though the bulk of Mexican corn was white, in the seed stores there were always a few ears with these mixed colours. When he talked to the local farmers he was told that a few striped seeds were planted out in each field as a charm, a fertility token, an offering of symbolic blood – though most denied that they ever indulged in the practice. *Sangre de Cristo* was one of the clues that helped Anderson to understand that the selection and propagation of new varieties of plants wasn't always guided by economic need in the first instance. It could be prompted by magical significance or community pride or pure faddishness.

Anderson was a field worker in the literal sense. He travelled across central America looking at maize types, field by field. In stretches of Mexico he would find that the character of the local corn changed dramatically even over a hundred miles. Often single fields would be devoted to one particular variety. Amongst the commonest were various kinds of popping corn. Far from being a modern invention, popcorn – *maíz reventador* in western Mexico, literally, 'corn which explodes' – is one of the earliest and most primitive forms. 'The ears,' Anderson writes of a typical specimen, 'were slender, almost the shape of a cigar, and not very much larger than one, their tiny white kernels set together like tiles in a pavement.'

Maize was hugely diverse even before industrial plant breeders got to work on it. The varieties that have appeared during its 10,000-year history are partly the product of the species' natural tendency to shape shift, and partly of early farmers' willingness to watch and attend to its changeableness. In 1983 Barbara McClintock won the Nobel Prize for physiology for her work on uncovering the mechanisms on genetic transposition, done largely by her intense observation of maize plants both under the microscope and in the field. She became so adept that she could eventually guess the chromosomal arrangements of corn varieties from their appearance, and add genetic 'footprints' to the evidence accumulating about the way maize moved around the Americas.

McClintock was a naturally intuitive scientist. After hours of gazing at chromosomes through a microscope she would sit under a eucalyptus tree and meditate, and find that solutions to their genetic geometry would often appear subconsciously. Her biographer, Evelyn Fox Keller, has described how her 'feeling for the organism' made links between her senses and her intellect:

> If this were a story of insight arrived at by reflection, it would be more familiar. Its real force is as a story of eyesight, and of the continuity between mind and eye ... Through years of intense and systematic observation and interpretation – she called it 'integrating what you saw' – McClintock had built a theoretical vision, a highly articulated image of the world within the cell. As she watched the corn plants grow, examined the patterns on the leaves and kernels, looked down the microscope at their chromosomal structure, she saw directly into their ordered world. The 'Book of Nature' was to be read simultaneously by the eyes of the body and those of the mind. The spots McClintock saw on the kernels of corn were ciphers in a text that, because of her understanding of their genetic meaning, she could read directly.

The work McClintock did on the variability of maize began to unstitch the idea that the genome of an organism was fixed. That parts of it could transpose spontaneously or because of environmental stress ('gene jumping' as it is now popularly known) helped start the study of epigenetics. The revelations of this new science have profound implications for the idea of plants as subjects of their own lives. They've partially vindicated, for instance, the contentious theories of the French biologist Jean-Baptiste Lamarck, who argued that characteristics acquired during an organism's lifetime can be passed on to its offspring. It also shows that ancient genes can lie dormant for epochs, even as they pass through different kinds of organism in the course of evolution, only to be switched back on when the need arises. McClintock believed her own discoveries had helped liberate the plant from the status of, in her words, 'a piece of plastic' to an 'object-as-subject'.

12

The Panacea: Ginseng

THE NUMBER OF PLANT REMEDIES used in orthodox medicine (they range from psyllium seeds for constipation to opium poppy derivatives for severe pain) is these days far exceeded by the range of 'alternative' herbal nostrums that have no scientifically proven effectiveness at all. The power of belief in healing is well understood by those who trade in potential treatments. Evaluating the way plants might influence disease, with its complex entanglements of mind and body, is a different kind of process from considering them as sources of food. Malaise and remedy aren't coupled as self-evidently as hunger and food. Filling an empty stomach is easy. Almost anything will do. If you have made a bad judgement about edibility, your body will quickly inform you of the mistake. Quieting a stomach ache is another matter, and the chains of cause and effect are far from obvious. By way of compensation the mind can play tricks with pain that it can't with hunger. Imagination and belief have played huge roles in healing systems where neither the disease nor the hopefully curative agents are properly understood. Intricate skeins of magical divination and mythic taxonomies have been spun to link disease and plant, and to make up for pharmacological deficiencies with legerdemain and persuasion. In the vulnerability of suffering, hope and hokum have always fed off each other. What is ironic is that in both indigenous societies, which value plants and believe in their 'spirits', and in pre- and post-scientific European communities plant-based medicinal systems have been uncompromisingly anthropocentric. Their common assumption is that plants don't cure human sickness by happy accident. It is the purpose of their existence on the earth.

But, in the margins, and over the past two centuries increasingly centre stage, the harder business of patient observation and meticulous trial and error have added a real-world corrective. The story of one plant, *Panax ginseng*, encapsulates the development of the medicinal plant's image, as it absorbed scraps of classical mythology and sympathetic magic, was rebranded by the Church, subjected to the more severe inquisitions of science, and then taken up by the consumer marketplace, the new generator of persuasive symbols. In the late 1960s and 70s ginseng became the most fashionable plant remedy to emerge from the decades' fascination with oriental traditions and mysticism and become a legendary cure-all. Its champions boasted that it would boost sexual performance, counter fatigue, aid memory and put your bodily and mental processes back 'in balance'. It became the late twentieth century's panacea, even though the majority of its users were quite unaware that this was, in a way, its original name.

That a genus of unprepossessing herbs from China and America was dubbed *Panax* by Carl Linnaeus in 1753 is of some significance. The name *Panax* derives from panacea, a classical name for a universal remedy. Originally, the term was used purely for medical remedies. Nowadays we talk of panaceas for everything from economic crises to adolescent alienation. Panacea was the daughter of Asclepius, the Roman god of healing, and the cure-all, early medicine's grail, was named after her. In the Middle Ages finding the panacea was one of the goals of alchemy, and there were many candidates. Classical authors such as Theophrastus applied the term Panax to various Syrian umbellifers, such as 'Hercules Woundwort or All-heal' (*Panax heracleum*). Other vaunted contenders were the 'Balsam of Fierabras' (Fierabras was a Spanish superhero from the time of Charlemagne), Prince Ahmed's Apple (a fruit purchased at Samarkand in the *Arabian Nights* tales), even the humble weed yarrow, *Achillea*. None of them worked as miracle drugs – though yarrow isn't bad as a styptic for small cuts.

When Linnaeus was revolutionising the classification and naming of plants in the mid eighteenth century, he decided to award the generic name of *Panax* to a group of quite different plants, related to the ivies. One species, *P. ginseng*, is a slow-growing perennial from South East Asia. The Chinese name *schinseng* means 'man shape', and refers to the distinctive forked and fleshy roots, with their phallic and limb-like outgrowths. It had been part of the Chinese *materia medica* for at least 2,000 years before it arrived in Europe in the seventeenth century. The Jesuit priest Athanasius Kircher was the first to refer to it in Western literature, in his *China Illustra* of 1667. In 1679 the Norwich physician and naturalist Sir Thomas Browne cited Kircher's botanical description when he reported that ginseng was being sold in London, though he never tried it himself, despite his famous and intractable melancholy. In the early eighteenth century a fuller and more heartfelt account was given by another Jesuit, Pierre Jartoux, who had been travelling as a missionary in China. He'd been given the tincture of a root to help ward off a spell of exhaustion, and was impressed by its effectiveness. He began to investigate ginseng's role in local culture, and found that it was then the most prized – and most expensive – medicinal plant in China. Its active part, the root, was used for a huge range of conditions. It was prescribed for cholera, colic, convulsions, depression, earache, fever, gonorrhoea, impotence, laziness, rheumatism and vertigo, but most commonly for fatigue and lack of sexual drive. There are steroid-like chemicals in the plant that have a direct, if mild, biochemical effect on some of these conditions. But the plant had suggestive and symbolic power too. The root's fanciful resemblance to the human form suggested it had been 'signed', by God or nature, as powerfully therapeutic for the whole body.

The idea that the physical form of a plant or its mode of existence in its own world are clues to its likely impact on the human body or spirit is one of the fundamental principles of pre- (and post-) scientific medicine. It is the part of the system known as sympathetic magic, which is often described simplistically as 'like cures like', but is really part of a more complex view of natural creation, in which all components are connected and have resonances with other components of a similar form or seasonal rhythm or position in a cosmological hierarchy. If you see the world as a connected whole, then exterior resemblances may indicate similarity in internal processes and effects. Sometimes sympathetic magic was indeed a simple case of 'like cures like': in Europe yellow flowers were given for jaundice, spotted leaves for rashes. But it could also be 'like *promotes* like' (red fruit to revivify 'tired' blood). In the complex mythology of the Tukano Indians in Colombian Amazonia, *uacú* trees (members of the pea family genus *Monopteryx*) have powerful sexual associations resulting from the V shapes between their buttress roots and their abundant sap. If the men decide to fell a *uacú* during clearing a patch of forest for cultivation, they build a platform around the tree, and cut the trunk with their axes. At a certain depth a sudden spurt of yellowish liquid will form a three-foot-long horizontal jet, and the men will jump under this shower, attributing it with the power to promote muscular strength and sexual potency. Sympathies can also be predicated on time or the seasons – night fevers being treated with night-flowering species, for instance, or by plants gathered in the dark. Eastern US Cherokees regard ferns as a sympathetic remedy for arthritis, because the young fronds are cramped, curled up, and gradually unfurl as the plant matures.

As well as the medicinal culture of its indigenous peoples, the American East Coast was a repository for Old World plant mythology and folklore. In the redoubts of the Appalachian mountains especially, herbal nostrums, folk recipes, fundamentalist beliefs and outright plant magic had been packed as cultural baggage with the cows and cooking pots by early settlers, and survived intact in isolated mountain communities well into the twentieth century. Southern Appalachia isn't a cultural backwater, a museum where the quaint folkways of old Europe are given a southern drawl; modern medicine increasingly works alongside traditional lore. But a core of ancient plant magic persists. The bark of the peach tree is still a folk remedy for diarrhoea. But the root of the tree must be scraped *upwards* in the gathering – against, so to speak, the direction of the runs. Similarly, before a woman went into labour, a sharp object 'was placed under the bed to "cut" (diminish) labor pains, and prevent or stop hemorrhaging. If an axe was used, some midwives insisted that it had to be one that had been used to cut down hundreds of trees, proving its power. It had to be placed under the bed sharp side up.'

In England, the cut tree itself was believed to have a directly sympathetic effect on healing. Gilbert White, in *The Natural History of Selborne* (1789), describes an extraordinary custom which had died out only in the previous century. (It persisted in Appalachia till the early twentieth century.) In the village of Selborne there was a row of pollarded ash trees which, by the long scars down their sides, had evidently once been split open. 'These trees,' White reported, 'when young and flexible, were severed and held open by wedges, [and] ruptured children, stripped naked, were pushed through the apertures, under a persuasion that, by such a process, the poor babes would be cured of their infirmity. As soon as the operation was over, the tree, in the suffering part, was plastered with loam, and carefully swathed up.' If the broken parts of the tree coalesced and healed, so would those in the child.

But in Europe during the sixteenth and seventeenth centuries, the principles and practices of sympathetic magic – sincerely believed in, if bizarre to our modern eyes – were refashioned into something more contrived, and known as the Doctrine of Signatures. In one respect this was another offshoot of the Genesis myth, and the arguments about what had happened in, and to, the Garden of Eden. Had there been diseases (and pests – also God's creatures) before the Fall? If so, had there been curative plants inside the Garden's walls? Or had disease – like agricultural toil and the pain of childbirth – been another of God's punishments dealt out to a disobedient humanity? It was widely believed that God would not have abandoned his children without at least the possibility of remedies – though these would have to be found with difficulty and with faith, otherwise the punishments were weightless. The difficulty was in identifying which plant was intended for which disease. The faith lay in a belief that God had 'signed' all plants with indications of their curative properties. The signatures just had to be 'seen' and understood. The Doctrine was a pharmacology based on decryption, a search for Intelligent Design in, so to speak, the plants' packaging motifs. The latitude this gave for ingenious personal interpretation was generous. Remedies for bad or aching teeth were seen, for example, in the tight-packed white seeds of pomegranate, the scales of fir cones and the ivory-coloured flowers of toothwort. As for ginseng, the manikin that could be glimpsed in the forked roots suggested it was a 'catholicon' for the whole body.

The epitome of Signaturism's convoluted thinking was the walnut. The nut – a bony shell containing a kernel lobed like the human brain – was given as a specific for disorders of the head. It hadn't developed to that perfection of form, Signaturists argued, for the sake of the tree, so that it could efficiently propagate its own kind, but for *us*, to remind us of our brains, and the superior understanding of the ways of the world God had planted there.

Many Signaturists' cures worked (and still work) for particular individuals, and their chosen plants must have acted powerfully on the placebo response. Just as modern pharmacologists know from their clinical research that the colour of a capsule can add to or diminish its physical effects, a plant whose leaves were shaped like a liver or a lung would have galvanised a patient's auto-suggestive, self-healing processes.

It's easy for us, with our privileged but far from complete scientific understanding of how curative plants work, to ridicule the propositions of sympathetic magic, whether it's practised in an Amazonian tribal community or an Appalachian trailer park. But they were a gesture, at least, towards the belief in a joined-up world, in which the components were connected either magically or physically. And how else, before the development of chemical analysis and monitored scientific trials, were genuinely curative plants to be found? One possible answer, usually overlooked, lies in the behaviour of our biological ancestors. Humans evolved from a 3-billion-year-old lineage of organisms which needed no conscious effort in deciding whether a plant would do them harm or promote their well-being. Those that chose badly died out, the others evolved, with a new piece of information stored in their genetic memories. Evolution by natural selection is trial and error in slow motion, and though we usually associate it with such physical features as body shape and beak sizes, it was also a process of development for instinctively recognising toxic and beneficial plants. Colour and pattern recognition would have been part of this discrimination, but volatile chemicals – the vaporous messages we call scents when we are consciously aware of them – were the most important mediators. A highly refined sense of smell has become redundant in modern humans, crowded out of brain space by the overwhelming use of visual imagery. But it persists in our animal relatives. Watch a wild herbivore, a deer, say, in a field or a garden, sniffing its way round the forage, choosing one species, ignoring another, avoiding the ragwort which is sometimes the downfall of domesticated stock, picking out pink tulips, chewing off buttercup flowers but not the more corrosive leaves. Plants producing fructose, fruit sugar, a significant source of energy, can be smelt by many terrestrial animals. It's no accident that most plant species whose fruits are both large and sweet are also non-toxic to mammals, so that their seeds can be safely ingested, excreted and spread about. Chimpanzees with infections have been observed sniffing out and eating plants which contain antibiotic chemicals. Gorillas in the Central African Republic rake through elephant dung for the seeds of jungle sop (*Anonidium mannii*) for its potent alkaloids, which seem to act as a sedative in digestive problems. The sensory systems through which animals analyse their environments are part of our genetic inheritance, for example in our intuitive wariness of plants which are black or smell of rotten meat or have acrid tastes. But it's a long step from these broad instinctive reflexes to deliberately choosing remedial plants for specific illnesses when the true nature of both may be obscure. Animals don't instinctively 'know' plant remedies for, say, heart failure or a new respiratory virus.

Beyond this there is the commonsense business of trial and error. It's a reasonable assumption that emergent humans added to their genetically inherited lexicon of food and healing plants by conscious testing. The popular myth envisions the process as a life-risking reality show, and history as littered with the corpses of tribal tasters who made a bad choice. Experimental ingestion of plants must have taken place at all stages of human development, with unpredictable results. But it would probably be a mistake to imagine this – at least before the eighteenth century – as a methodical or even deliberate business, a case of 'let's try this plant to see if it affects this condition'. Accidental discoveries are more credible. The fruits of common buckthorn, for example, are powerfully purgative, and were widely used as such at least as far back as the Middle Ages. When the latrine pits of the Benedictine Abbey at St Albans were excavated, quantities of buckthorn seeds were found stuck to old rags the monks used to wipe their bottoms. The fruits are simple black berries in which even the most imaginative Signaturist would have been hard pressed to glimpse any cryptic indications, and doubtless their purgative powers were discovered much earlier, in the traumatic aftermath of an ill-advised fruit supper.

There is another possible route to therapeutic revelation. In many indigenous communities shamans claim to be able to identify curative plants during trance states. Trances can be induced by chanting and rhythmic dancing, but in Amazonia the most used and respected route to knowledge is via drugs, especially the potent mixture of hallucino-genic herbs known as *ayahuasca*. The two commonest components of this are leaves and twigs from the vine *Banisteriopsis caapi*, which contain chemicals similar to monoamine oxidase inhibitor (MAOI) anti-depressants, and the coffee family shrub *Psychotria viridis*, whose active component is the powerfully psychoactive dimethyltryptamine (DMT, a controlled drug in Europe and America). In *ayahuasca* rituals, brews of these plants are drunk communally, the ceremony being guided by an attendant shaman, who also helps interpret the experience. An inevitable but psychologically important prelude to the visionary stages is the copious vomiting that follows soon after drinking. This is seen as a necessary purification of both mind and body. The consciousness altering follows in an hour or two. Western 'spiritual tourists', with their particular expectations, tend to describe their hallucinations in dramatic but nebulous terms. They talk of a sense of rebirth, or glimpsing 'the meaning of the universe', or of a personal revelation – of finding, as one experimenter told me, 'a place of safety I could always return to'. Anthropologists who have studied the reactions of indigenous people find their descriptions less personalised and more grounded. They never have visions of 'strange beings, monsters or alien landscapes; there is nothing new, so to speak. They see the land, peopled by recognisable ancestors, the well known animals, the familiar trees and rivers. The difference from ordinary reality is that the dead now speak and admonish, teach dances and songs, spells and cures.' Gerardo Reichel-Dolmatoff, who has spent much of his life working with the Tukano Indians in north-west Amazonia, explains how the shaman interprets the visions of medically active plants. During trances plants and animals

> tell the visionary how they want to be treated and protected so they can better serve him; how they suffer from carelessness, over hunting, the cutting down of trees, the abuse of fish poisons, the destructiveness of firearms. Seen from this perspective we must admit that a *Banisteriopsis* trance, manipulated by shamans, is a lesson in ecology, in the sense that it gives nature a chance to voice its complaints and demands in unmistakable terms. Since everything seen and heard in the trance state is already known from traditional shamanic teaching, the trance only proves that shamans had been right all the time when they said that the ancestors, the plants and the animals, the forest and the river, were a living presence.

What anthropologists portray in their accounts of the structure and workings of these indigenous communities tends not be the hierarchy-free utopia imagined by some Westerners. Amongst the Runa people in Ecuadorian Amazonia, Eduardo Kohn found a kind of spirit feudalism, in which the spirit masters are envisioned as large-scale farmers of the forest, who for their part, view the wild birds, animals and plants as their domestic poultry, hunting dogs and food crops. The Tukano, though deeply and ecologically aware of their environment, imagine it as an extension of themselves. They don't perceive the universe as a living organism, Reichel-Dolmatoff explains, rather, that humans participate in the cosmos and their immediate forest environment through an energy circuit 'which includes all plants and animals, together with all sense data ... But the universe or our earth are not thought to be alive as a system as such. What gives it life are humans incorporating what we call nature, into the human scale' – a belief not far removed from that of some European philosophers and theologians in the seventeenth and eighteenth centuries.

How the shamans themselves attain their knowledge of medicinal plants is a more testing question. Their own answer is that the plants 'speak to them', or at least inform them directly, though just how is never made clear. In the much-touted book *The Cosmic Serpent* (1998) one-time Stanford University anthropologist Jeremy Narby elaborated an extraordinary theory, that through the revelations of *ayahuasca* shamans are able to make direct spiritual contact with the DNA and genes in plants, and thus divine their pharmaceutical effects. DMT has been popularly called 'the spirit molecule' for its seeming ability to give the human brain the simultaneous powers of telepathic communication with plants and the penetrating vision of an electron microscope.

Shamanism is a cryptic and complex business, inseparable from the cosmologies of pre-industrial peoples, and does not translate easily into the thought structures of modern rationalists: if you're sceptical about humans' possession of souls, you will find it even harder to believe that plants have 'spirits' and wish to communicate their beneficence to us. Paranormal explanations such as Narby's dreamed-up theory do no service in helping to communicate the objective workings of indigenous plant medicine. A high proportion of Amazonian plants contain physiologically active and toxic compounds, and it's no surprise that local inhabitants have discovered a good number, despite having arrived in South America from Asia maybe as recently as 15,000 years ago. Their knowledge of curative plants for parasites and psychosomatic disorders is especially refined. On the other hand, they aren't omniscient, and have proved powerless to deal with contagious diseases introduced by Europeans colonists, such as measles, smallpox and syphilis, which decimated South American native populations.

Reichel-Dolmatoff suggests more down-to-earth routes to plant knowledge. *Ayahuasca* makes no direct contribution to this. The drug, with its uniform effects underlined by the shaman, seems chiefly to be

A shamanic portrait of a Korean mountain deity, Sansin, holding a root of wild ginseng.

employed as a way of enhancing social harmony and promoting shared values, especially the etiquette of respect for the forest. But it may have a role in enabling cognitive short cuts, based on a deep sensory appreciation of plants and their interdependent animals. This echoes the intuitive process – 'a feeling for the organism' – used by Barbara McClintock after long hours communing with maize plants through an optical microscope, 'a story of eyesight, and of the continuity of mind and eye'. The Tukano can apparently cure themselves of minor illnesses by simply entering the forest, while concentrating acutely on sounds, odour, colours, the behaviour of insects and the temperatures of different layers of vegetation. Since, in Dolmatoff's words, 'the literal and the metaphorical are inseparable in their world-view' they are able to see 'the forest [as] a memory device in which all sensorial perceptions are registered and trigger associations, awaken memories which help solve personal conflicts'. He describes one example of how symptoms of physical fatigue combined with claustrophobic agitation (perhaps the beginnings of a migraine?) are imagined by the sufferer as a basket, woven tightly from twigs, which is enclosing the sufferer's head. By entering the forest – 'a huge basket full of everything' – the lesser basket is humbled, and the patient recovers. This is classic sympathetic magic, but informed and made coherent by an intense awareness of the physical details of the environment.

In the West, cognitive behaviour therapists treat tension headaches by coaching sufferers in imagining, then relaxing, the 'tight bands' round their head. Four hundred years earlier their professional forebears would have recommended doses of walnut. For better or for worse, across centuries and cultures, metaphorical images of plants and vegetation have been a fundamental ingredient in their power to heal.

Five thousand miles north of Colombia, Native American peoples also had theories of disease. The Cherokee believed that humans were free of illness until the animals created them in retribution for the lack of respect humans had shown to nature – a pagan version of the punishments of the Fall. But the plants felt the animals had been too harsh, and volunteered themselves to provide a cure for all the illnesses the animals had created (a myth obviously created before plants too began to be shown similar disrespect). In Canada, one of the remedies recommended by Iroquois medicine men was the root of a small and unassuming plant, rarely more than sixteen inches, which grows right down the eastern side of North America, from Quebec and Manitoba to Alabama and Arkansas. This is American ginseng, *Panax quinquefolius*, a species which clearly shows the ginsengs' kinship with the ivies. It likes shady deciduous woods on rich soils, puts out greenish yellow flowers followed by clusters of red berries reminiscent of the small fists of ivy fruits. The manikin roots take eight years to form and usually have more limbs than an octopus. The Cherokees have used it for much the same range of complaints and inadequacies as the South East Asians use their species.

When news of the oriental panacea first reached Europe in the eighteenth century, a new market for the plant opened up. The reputed boost it gave to sexual potency was its most marketable feature, and demand grew to such an extent in Europe that the price for a perfectly, suggestively shaped root rose to ten times that of the same weight of gold. Entrepreneurs began to look out for other sources of the plant, for home consumption and also for export to China, where stocks were in increasingly short supply.

The first written evidence that Europeans had cottoned on to Native Americans' ginseng habit is in the correspondence of settlers in the late seventeenth century. They kept the herb for what they prudently called 'private use'. The Virginian planter William Byrd wrote a titillating account of his morning cup of ginseng tea, made just as the Chinese did, by simmering the root in a silver pot over a charcoal fire:

> It gives an uncommon Warmth and Vigour to the Blood, and frisks the Spirits beyond any other Cordial. It chears the Heart even of a Man that has a bad wife, and makes him look down with great Composure on the crosses of the World ... In one Word, it will make a Man live a great while, and very well while he does live ... However 'tis of little use

in the Feats of Love, as a great prince once found, who hearing of its invigorating Quality, sent as far as China for some of it, though his ladys could not boast any Advantage thereby.

Again it was a Jesuit, Fr Martineau, working as a missionary in what was then French Canada, who made the benefits of the American species public, and soon the French were shipping large quantities direct to China, using Native Americans as collectors.

The harvesting of American ginseng for the foreign market spread south to Byrd's part of the world, and what was left of the Appalachian forests. During the depression of 1857–8, when many small farmers went bankrupt, it became an invaluable wild cash crop. Whole communities turned to hunting the herb, a custom that came to be called 'sanging'. The town of Ginseng, in La Rue County, Kentucky, is named for the herb, which was sold at market in nearby Elizabethtown, though there's not much evidence of 'home use' by the picking communities: this was a commodity for export. But though it helped tide the local economy over, in many places ginseng was overpicked. The decline was accelerated by the increasing destruction of the Appalachian forests for coal and mineral mining, and some first attempts were made at cultivation techniques which had been introduced in Korea with the Asian species.

They weren't too successful to begin with. American ginseng is a sensitive, old forest species, shade loving and fussy about soil type. It didn't take well to being grown in fields and backyards. More recently forest-farming techniques have been introduced, with chosen strains being planted out – and minimally tended – in the forest itself, which produces higher yields but makes the crop vulnerable to thieves. The spur for the revival of ginseng's fortunes was its sudden rise to fame as a herbal stimulant in the countercultural mood of the late 1960s. Body builders, sexual adventurers, the perennially fatigued and anyone in search of a novel high began to take it. And, as with historical panaceas (and modern 'super nutrients'), its aura of potency began to spread beyond its original focus of action. Soon ginseng was being added to shampoos, skin creams, soft drinks and vitamin supplements. All these products are available over the counter, and the new demand (though it has declined now) revitalised the ginseng trade. The vast majority of the world's ginseng still finds its way to China, where it is used almost universally as a tonic, and where a perfectly man-shaped root can still fetch up to $10,000. Ginseng picked in America is still sent to the Far East for wholesale trading, before finding its way to a Chinese banker's bedroom or back to a US drugstore.

In 2000, 300 tonnes of American ginseng was exported to Hong Kong, and 'sanging' remains an important part of local Appalachian economies. At the point of highest demand in the mid 1990s almost 100,000 pounds were gathered annually in the states of Kentucky, West Virginia and Tennessee alone. Collectors receive an average of $500 a pound – which makes ginseng the most valuable plant crop in the United States.

Whether the panacea works is questionable, at least by Western medical criteria. Accounts of its use and effectiveness appear to vary according to the culture in which it's being tested, suggesting that social values and expectations are powerful psychological boosters. In the USSR, Chinese ginseng (or more probably the unrelated Siberian 'ginseng' *Eleutherococcus senticosus*) was given to cosmonauts to increase their stamina, and to factory workers to increase their contribution to the Soviet GNP. During the Vietnam War the Viet Cong used ginseng to treat gunshot wounds. Orthodox Western medicine has tested ginseng and found it has few measurable effects beyond an occasional rise in blood pressure and marginal increase in stamina. In London, nurses on night shift reported feeling more alert on duty when dosed with ginseng, and in a more stringent but typically unpleasant laboratory test, rats given the herb took longer to drown when forced to swim in an escape-proof tank. In the USA in the 1960s, the godfather of modern foraging, Euell Gibbons, issued a laconic verdict on the herb's effects. He brewed up some tea from roots he'd gathered in a Pennsylvania wood, and sipped it as pensively as William Byrd. He detected no physical effects what-soever, but confessed that 'the feeling of luxurious self-indulgence that came from drinking a beverage that would have cost a Chinese a king's ransom was terrific'. Rational hedonist to his marrow, he thought a tincture might make a novel cocktail bitter.

Practitioners of complementary medicine have invented the term 'adaptogen' to describe ginseng's elusive effects on human physiology. They argue that it helps the body adapt to stress of all kinds, normalising energy levels, immune responses, appetite and mood – which is why its effects are so hard to quantify. It sounds like an easy way of avoiding the discipline of serious testing, but it might be true. Such chemicals are well known in the plant world. One is salicylic acid – the precursor of aspirin. It was discovered in willow, but is widespread in plants. It acts botanically as a hormone, promoting growth, reducing the effects of stress and, in the event of damage occurring to one part of a plant, carrying messages to boost resistance in other parts. Its action on humans isn't identical. We use its synthetic derivative aspirin (named incidentally after another botanical source of salicylic acid, *Spiraea* (now *Filipendula*) *ulmaria*, meadowsweet) most often for pain, which isn't experienced consciously by plants, though damage causes electrical storms in their tissues. What is intriguing are the new revelations about aspirin's effect on the human immune system and its slowing of the growth of many human cancers, which echo its action inside plants, and remind us that our evolutionary roots lie in common primeval cells. Because of this shared ancestry there is always a chance of finding vegetal compounds which will be therapeutic in humans; we are simply dipping into the common legacy of biological self-medication. Antibiotics, for example, have been derived from the chemicals plants use to ward off fungal and bacterial infections. Astringents such as tannins, which shrink tissue and help heal wounds in humans, fulfil a similar function in plants, as well as warding off predatory insects.

But we're not plants. Our bodies have physiologies which are unique to our species, and to the animal world. Many compounds which are beneficial to plants can be toxic in our systems. So can those which are probably waste products, or a kind of incidental chemical ornamentation – for example atropine, from deadly nightshade, which in small quantities is used to dilate the pupil in eye examinations, but is lethal in high concentrations. The belief that 'out there, is a cure for everything' is yet another offshoot of the persistent human conviction that we are the focal point of all biological activity. Plants evolved their remarkable chemicals for their own purposes, something demonstrated by the precision of a 'magic bullet' chemical in the lima bean. If the bean is attacked by spider mites it gives off a volatile pheromone which attracts another species of predatory mite which feeds on the original attacker. But not any random predator: the bean analyses the spider mites' saliva and releases a volatile chemical which 'calls' only the predator species that feeds on that particular mite. It's conceivable that, by pure coincidence, this pheromone might activate human immune system cells thus fulfilling the teachings of sympathetic magic and analogy. But it is vanishingly unlikely in reality. Panaceas for all our human ills will not automatically be waiting in plants' unexplored depths.

13

The Vegetable Mudfish: Samphire

I MADE THE ACQUAINTANCE OF SAMPHIRE on the North Norfolk coast in the 1960s. It's a local delicacy, pickled, or cooked like asparagus, or just munched raw in the creeks. It looks unprepossessing, a bunch of floppy green strings, a spine-free marine cactus with its succulent segments sporting barely visible scales in place of conventional leaves. But my first encounter with it challenged the preconceptions I had about plants, the forms they might exist in, the bizarreness of their lives, their possible feral dangers. That primal sprig was the vegetable equivalent of a mudfish, a marine organism making the supreme adventure of colonising the land. It had flowers, minute dots of yellow on the fleshy stems, but they spent half their lives blooming under the sea. When I first tasted it the iodine-and-ozone tang was unlike anything I had experienced in my mouth before. Even in this basic role as an eccentric comestible, it seemed transgressive, a chimera from a medieval bestiary, half vegetable, half miniature sea monster. On the occasions it was sold commercially it was in fishmongers' shops or on quayside stalls, not by greengrocers. Huge and eccentric specimens were sometimes mounted above pub bars, as if they were prize pike. I once saw a colony that had invaded an abandoned dinghy, taking over the silted deck one low tide then rising up at high water to become a waterborne buffet counter.

Our gang picked our own in the muddy creeks, where the sheaves of branched green shoots could grow up to a foot tall. These foraging trips were like baptisms of mud, submersions in an element where life had different mores. Shoreline mud isn't like the mud in a cart track. It's gelatinous, shiny, grasping. It's a terrain for fleeting, cautious visitors – the wickering redshank that flew about us, the scuttling crabs, even ourselves, because though we sometimes thought we would be sucked irretrievably into its warm gloopiness, we always managed to scramble clear. But not, surely, for *plants*, the epitomes of fixedness. Later I saw samphire on the intertidal mudflats, in another kind of growth form, its young green spikes tightly covering acres of the glutinous surface. It's an annual plant, the first urgent coloniser of bare mud, and it made these bare shoals at the edge of sea look like bowling greens. We were instructed in harvesting lore by a local called Crow, a jack of all coastal trades. His stern picking rules seemed to link social tradition and ecological imperative. You should never wash or store 'sea asparagus' (pronounced 'sparrowgrass' locally) in freshwater, which would suck out its sap, leaving only a wilted shell. What you have picked should have been 'washed by every tide'. The latter seemed a sensible laundering principle, given the execrable muddiness of its habitat, but also evidence of astonishing endurance for a flowering plant. Samphire's adaptation to saltwater dousing is the same as a desert plant's to drought. Both habitats lack fresh water, so plants store their own in 'aqueous tissue'. They become succulents. What helps the plant survive is what causes it to melt like concentrated green jelly between the teeth.

But there seemed something counter-intuitive about its life cycle. Its ability to survive submersion by the tides, twice daily for six months of the year, means that samphire creates the conditions for its own annihilation. The stems on the mudflats closest to the sea are densely packed, and catch debris washed in by the tides – grains of sand, flecks of mud, crushed shells, dead shrimps, seabirds' feathers – and hold them together as silt, which is cemented further by insinuating growths of algae and bacteria. Samphire *builds land*. I'd watch the flats for hours, seeing a whole embryonic ecosystem rise and sink. Sometimes a tidal surge would breach the sea-wall defences, and change the whole order of things. Liquid mud would flow into hollows in the shingle, or deep into the freshwater grazing marshes. Samphire would smartly follow, doing exactly what it has evolved to do. A diligent botanist in the 1930s,

First footing. Samphire colonising bare mud in the Wash, England.

probing the mudflats at Scolt Head just a few miles west of where we foraged ourselves, had found that, in the communities of samphire at the lowest edge of the saltmarsh, the land was rising at the rate of a quarter of an inch a year. In fifty years (assuming no savage tidal surges) it would have risen by just over a foot and dried out enough for perennial plants like sea aster and sea lavender to become established, and samphire able to find ever fewer open patches for its seeds. In 200 years the surface of the marsh would theoretically have risen by more than a yard, and might not be washed by any tide, let alone two a day. Samphire, simply living the way it does, makes its *terroir* untenable for itself. It seemed to me, then, to fly against every principle of evolution as I understood it. Why hadn't a super samphire developed by natural selection, a strain which grew more widely spaced, or whose root system self-destructed just before seeding, so that the mud was left in the deliciously oozy fluidity that the plant was best adapted to?

I eventually learned that this wasn't how evolution worked outside the laboratory. Plants might be overridingly concerned with the survival of their own genes, but except in the case of a very few forest trees, none of them manipulate their habitats to ensure their own continuance over other species. That is our own species' dubious prerogative. Nature abhors vacuous monocultures, with all their intrinsic vulnerabilities, and in the real world evolution is a social business, where diversity and succession and give and take are the rules. Plants don't have 'purposes', but they do have roles, and fulfil these in specialised situations and often for strictly limited time spans. A samphire lawn may become a grass pasture dry enough to graze cattle, but one good storm surge and the vegetal mudfish will be back, taking advantage of solid land that has become mobile again, and slowly reclaiming it.

Some years on I saw a patch of flood-threatened East Anglian coastland where this had been allowed to happen deliberately. The sea walls, ineffective barriers to storm surges, had been bulldozed down round a patch of farmland, to see if a naturally formed saltmarsh would be a more effective buffer. The first storm washed the cultivated soil away, but within a year samphire was invading the bare patches, building a different and more absorbent land, capable of soaking up both the sea water itself and its furious energy. Inviting, so to speak, vegetation to suggest its own solution to environmental challenges is different from treating it as a submissive service provider, and, I'd suggest, a better basis for a long-term relationship.

I wrote my first full-length book, *Food for Free*, a kind of post-modern guide to foraging, prompted by this quirky sea vegetable. Samphire taught me that the divisions between plant as sustenance for the body and nourishment for the imagination, and between scientific fascination and Romantic inspiration, were fluid. Everything I have written since has been influenced by the thought of its ephemeral life, in which opportunism, self-expression and utility are able to coexist.

THE SHOCK OF THE REAL: SCIENTISTS AND ROMANTICS

AS THE WAR CLOUDS THICKENED in 1939, the historic collections of dried and mounted plants prepared by Carl Linnaeus in the eighteenth century and stored at the Linnaean Society headquarters in London were moved for safety to Woburn Abbey in Bedfordshire. The following year, as an insurance policy, a photographic record was made of these systematically arranged herbarium sheets, in which the great classifier's view of botanical order had found its physical expression. In the process of posing the stinging nettle sheet for its portrait, the photographer, Gladys Brown, was stung on the arm, 'raising a definite blister apparently similar to one produced by a fresh specimen'. Two hundred years old, and as desiccated as a skeleton, the nettle had bitten back.

It would be easy to make a parable out of the stinger's dogged refusal to be emasculated. Linnaeus's 'system' was one of the bedrocks of biological science in the eighteenth century, but deeply nettling to poets and Romantics of all sorts. His invention of the binomial system, in which all organisms could be named by just two terms – the first identifying the family and the second the species – revolutionised taxonomy, and therefore biology. But he couched it in a foreign tongue, the Latin of the educated elite, and alienated those who saw nature as a commonwealth. His further attempts to classify and name plant species according to the numbers and disposition of their sexual parts (and describe these as if they were bohemian human liaisons) was a piece of comparatively short-lived whimsy, shocking to decent folk though surviving into the Victorian era as a colourful way of identifying flowers for an album. However, it ignored the more complex and subtle ways in which plants were related to each other and to their fellow organisms, and was soon superseded.

But the contemporary hostility to Linnaeus was deeper and more idealistic than an objection to this or that criteria. It was the idea of universal ordering itself that raised the Romantics' hackles. They saw his 'naming of the beasts' and building of biological family trees as Adamic hubris. With plants, it implied that they were fixed, predictable entities, devoid of vitality. The poet John Clare felt an especial bitterness about what he called the 'dark system', which had removed plants from their proper homes, and in its Latinate language had stolen them from common understanding too. 'I love to see the nightingale in its hazel retreat,' he wrote in his journal, '& the Cuckoo hiding in its solitudes of oaken foliage & not examine their carcases in glass cases yet naturalists & botanists seem to have no taste for this poetical feeling they merely make collections of dryd specimens classing them after Leanius into tribes and families ... I have none of this curiosity.' I've written about Clare's feeling for plants elsewhere, and he has been a powerful influence on my own thinking and that of an increasing range of modern writers. He was not as antithetical to science as he liked to pretend, and was skilled enough in discriminate observation to make the first records of more than forty plant species for his home county of Northamptonshire. But he was uncompromising in his feeling for them as subjects, not objects in collections. Not even Coleridge or Wordsworth in their most animistic moods would have felt able to begin a spring poem with such open arms as Clare. 'Welcome, old matey!' he exclaims to an April daisy. Or to compose an entire poem as if it were a plea by an exploited patch of vegetation ('The Lament of Swordy Well').

There was a powerful contemporary symbolism in the 'dryd specimen', evidence of the pressing out of life, the expulsion of the vital force that made living plants different from stones. The tone of much of post-Newtonian science was reductionist, intent on imposing order on the apparent anarchy of nature, and explaining its organisms and processes according to mathematical and mechanical laws. Not everyone approved. In 1817, at a very lively party at the painter Benjamin Haydon's house, Wordsworth, Keats and Charles Lamb had drunk a toast to 'Newton's health and confusion to mathematics!' Keats raised his now famous objection to Newton, that he 'had destroyed all the Poetry of the rainbow, by reducing it to a prism'. The idea that understanding could be seen as destructive of poetic feeling is hard for us to grasp today, comparable to suggesting that knowing the tonic scale destroys the possibility of appreciating the beauty of music. Yet most of the Romantics retained some kind of religious belief, and the deconstruction of big phenomena – light, plant growth, life itself – seemed to trespass into territory that was the prerogative of a Creator. And if all was explained, where was the space for the inherent ambiguity of poetry?

The rainbow wasn't the best Romantic stick to beat Newton with. It was already 'reduced' – or 'unweav'd', as Keats later put it – from sunlight, created in nature itself through the refractive properties of raindrops. A more apt target would have been Newton's Second Law of Thermodynamics, with its gloomy declaration that, in all closed systems, entropy (or disorder, put crudely) is inexorably on the increase. A perpetual-motion machine is an impossibility. There can be no sudden replenishments of energy from out of the blue. The workings of the universe are remorselessly running down.

The discovery of extraordinary new species and vegetal processes during the late eighteenth and nineteenth centuries tended to suggest otherwise, that the plant world was constantly renewing and expanding itself, and the spirits of those who engaged with it. Most scientists, philosophers and creative artists shared this optimistic vision. But their different perspectives on how plants 'fitted' – as machines, or property, or conduits for some nebulous creative energy – generated the excited tension that permeated the contemporary debate about the vegetal world.

14

Life versus Entropy: Newton's Apple

NEWTON'S BEST-KNOWN FLIRTATION with the vegetal world was to witness a falling apple and be inspired to posit gravity as a fundamental cosmic force consistent with his Second Law of Thermodynamics. Both laws insisted that objects could not spontaneously fly upwards from the earth's surface. Newton was living at the family farm at Woolsthorpe Manor in Lincolnshire at the time the idea struck him, and practising science as a cottage industry, as it was to be practised by 'natural philosophers' for the next two centuries. His study remains much as it must have looked in the 1660s, with one window reduced to the slit he used for directing sunbeams through his prisms, and the perennially sensible presence of a bed. Outside there are still remnants of the working farm, and of the tree that deposited the historic apple. It doesn't have its original trunk, but it is still the same organism, sprouted anew from the withered stump in a defiant challenge to entropy. Its fruits are big and heavy, and would certainly have made an impression had one fallen on Newton's head, as in the legend.

What actually happened is related by the antiquarian William Stukeley, who had dinner with Newton in 1726, not long before his death, and listened to his recollections of an autumn day at Woolsthorpe:

> After dinner, the weather being warm, we went into the garden and drank there, under the shade of some apple trees … he told me he was just in the same situation, as when formerly [probably in 1666], the notion of gravitation came into his mind. It was occasioned by the fall of

an apple, as he sat in contemplative mood. Why should the apple always descend perpendicularly to the ground, thought he to himself ...

To an eighteenth-century botanist, an equally perplexing question was how an apple could, as it were, be raised perpendicularly *from* the ground, how biological growth could defy gravity. What was the vital force that made life able to challenge the Second Law's vision of an ever descending spiral of energy? The peculiarity of Newton's apple – it's now recognised as a scarce variety called Beauty of Kent – added a kind of lateral assault on the Law, contradicting the gravitas of Linnaean certainties and the idea of 'species fixity'. By the eighteenth century there were tens of thousands of apple varieties in existence, all the Old World varieties at least now known to have descended from a single species in central Asia in a glorious pan-continental proliferation. The generation of biological forms (what we call biodiversity today) and the tendency of all living systems to become progressively more diverse and complex fly against the cosmic gloom of the Second Law. The polymathic biologist and essayist Lewis Thomas put the energy equation represented by biology's intervention into the inanimate world quite simply. '[T]he information of the biosphere,' he wrote, 'arriving as elementary units in the stream of solar photons ... [is] rearranged against randomness.' Life – at least until the stars burn out – trumps entropy.

That the domestic apple originated from a wild apple species in Tien Shan in north-western China had been suspected by the pioneering Russian plant geographer Nikolai Vavilov back in the 1920s. For a long while it had been assumed in botanical and horticultural circles that the domestic fruit had the European crab apple, *Malus sylvestris*, somewhere in its pedigree, either as a direct ancestor or a contributor to the lineage. Exhaustive studies on the DNA of domestic varieties have proved this assumption to be wrong. All domestic varieties are basically the same species, *Malus pumila*. Controlled attempts to cross crab apples with other *Malus* species and varieties have all been literally fruitless.

At the start of the twenty-first century the distinguished Oxford botanist Barrie Juniper, working both in the laboratory and in the field in Tien Shan, pieced together for the first time a full and plausible outline of the evolution of the domestic apple. It is a complex story, involving turbulent geography, genetic effusiveness, improbable mammalian fruit gourmets and prehistoric orcharding experiments.

The topography of Tien Shan was a crucial factor. For much of its history it was subject to violent earth movements, bringing ancient rocks of every geological type to the surface, creating new canyons and cliffs and caves, exposing fresh soil and breaking up existing vegetation cover. As a result the region became both very fertile and full of botanical refuges, cut off from neighbouring areas. It developed 'fruit forests' in which the precursors of many important species grew: apricot, almond, plum, pear, quince. At some point possibly 10 million years ago, the fruits of an unknown form of apple – maybe one related to the Siberian crab, *Malus baccata* – were carried by birds into this ecological labyrinth. Different populations were isolated in different valleys, and would have evolved through spontaneous mutation, and selection pressure from soil and habitat variation. Occasionally these populations would have come together and interbred, and the region would have begun to support an ur-apple species close to today's *Malus pumila*, intrinsically and epigenetically varied in size, shape, colour, sweetness and time of fruiting. The main drivers of selection, the pre-human apple breeders, were probably bears, which were common in the region thanks to the abundance of caves as secure refuges, and the similar abundance of fruit. Bears browse on windfalls and climb trees for the choicest fruit.

Chinese brown bears would have selected the sweetest and largest apples, scattering and manuring their pips with their faeces. Apples don't come true from seed (that is the corollary of the immense variation in the fruits) but the bears' choices would have shifted the gene pool in the direction of greater size and palatability. They were the midwives of the modern apple. Wild horses were another important vector. They also like apples and would have added their own well-dunged contributions to the seed bank.

Some 7,000 years ago nomadic human tribes began settling seasonally in the fruit forests, and added their own selection to that already achieved by wild animals. Maybe favourite trees were manured, and the virtues of pruning discovered when trees were broken or lopped for firewood. As tribes migrated west along the Turkic Corridor they would have taken fruit with them, and wilding apples would have followed, springing in unpredictable diversity from discarded cores and the dung of pack horses. Roughly 4,000 years ago apple grafting was discovered, inspired perhaps by the glimpse of a natural pleach between two chafing branches. From this point it became possible to perpetuate favoured varieties by surgically implanting a slip onto another apple stock. It's not at all impossible that an apple with some of the characteristics of Newton's Beauty of Kent had already appeared, and been enjoyed enough to be passed on. From this moment the voyage of the apple joins and resembles that of other cultivated crops.

Throughout the Middle Ages apple varieties were brought to Britain from the continent, or spotted as rogue seedlings in existing orchards. John Parkinson made a euphonious list in his *Paradisus* of 1629, including:

> The Gruntlin is somewhat a long apple, smaller at the crowne than at the stalke, and is a reasonable good apple.

> The gray Costard is a good great apple, somewhat whitish on the outside, and abideth the winter.

> The Belle boon of two sorts winter and summer, both of them good apples, and a fair fruit to look on, being yellow and of a mean bignesse.

> The Cowsnout is no very good fruit.

> The Cats head apple tooke the name of the likenesse, and is a reasonable good apple and great.

There is no mention yet of the Beauty of Kent, though there is a Flower of Kent, 'a faire yellowish greene apple both good and great'. Nor is it listed, a hundred years later, in Philip Miller's *Gardener's Dictionary* (1732), though his catalogue confirms the continuing expression of *Malus pumila*'s genetic inventiveness in every dimension from scent to fruiting season. (It includes the aniseed-scented 'La Fenouillet', and 'Le Courpendu ... or the Hanging Body' also known then as the 'Wise Apple' because it flowers late and escapes spring frosts.)

The global partnership between fruit growers and fruit continued into the nineteenth century. In the United States (where American wild apple species were brought into the breeding line) Henry Thoreau added a Romantic gloss to the roll call by describing mood-and-moment varieties: 'the Truants' Apple (*Malus cessatoris*), which no boy will ever go by without knocking off some ... our Particular Apple not to be found in any catalogue, Malus pedestrium-solatium ...' In Normandy in France, a group of highly local cider apples were developed. And in 1841 Thomas Squire, who lived in the town where I grew up, Berkhamsted in Hertford-shire, was planting out a curious *Malus* seedling just as Queen Victoria and Prince Albert were making regal progress past his garden towards the Kings Arms, my own boyhood local. He named it 'Victoria and Albert', later to become 'Lane's Prince Albert' after the nurseryman, John Lane, who successfully commercialised it. But the tree survived in Mr Squire's garden until 1958, and the fruit still holds a Thoreauvian aura for me – 'The Apple of carefree youth, the Sunday lunchtime apple ...'

The second half of the nineteenth century probably marked the zenith of apple diversity. Twenty thousand varieties are believed to have existed worldwide, with 6,000 in Britain alone. The Welsh–English borderlands had one of the densest concentrations of apple orchards, but they were beginning to be abandoned because of the prolonged agricultural depression. This was the home country of the enterprising Woolhope Naturalists' Field Club, based in Hereford and commanded by the tireless Dr Henry Graves Bull. Dr Bull was the originator of the Fungus Foray, an annual four-day event in which members of the society toured the countryside in coaches, inspecting and gathering fungi which were later served up to them at a grand dinner in the George and

Dragon public house in Hereford. The forays took in the local orchards, and members were able to see at first hand the decrepit state many had fallen into. In the early 1870s Dr Bull suggested that the club take up the subject of 'the Pomology of the county', and investigate with some urgency the history and surviving variety of its apples and pears. The Royal Horticultural Society's head of gardens, Dr Hogg, was enlisted as technical advisor, and a series of exhibitions was held in Hereford, to which members and local fruit growers sent samples for naming. Many couldn't be named 'for they had no names ... [but] were valuable apples, so quite distinct in character, and with such excellent qualities that they deserved to be better known'. Dr Hogg suggested that a local 'Pomona' (an illustrated catalogue of apples) should be compiled, and offered to edit it himself.

The first part, in quarto format and with watercolour illustrations, appeared in 1878. Herefordshire proved to be a hot spot of pomological diversity. Twenty-two species were included in the first tranche, rising to forty-one in the second, published in 1879. The project proliferated much like *Malus* itself, infiltrating new regions, absorbing local skills, developing a special vocabulary of fruit description. In 1880, in response to widespread acclaim for the first three parts, the club decided to make the Pomona a national project. The autumn exhibitions became more lavish and imported specimens from other shows. That October there were more than 2,000 dishes of fruit on display. In 1884 club members crossed the Channel to visit, and exhibit at, an apple show in Rouen in Normandy, the organisers of which proved to be compiling their own cider apple directory. This provided more revelations about the radiation of apple variety: what, in Herefordshire, had been called 'Norman' cider apples proved to be quite unlike any grown in Normandy. Specimens of the true Norman apples were accordingly taken back to England where, 'in some haste', paintings and descriptions of them were made and added to the Pomona.

The work eventually ran to seven parts, which were bound into two volumes and published as *The Herefordshire Pomona* in 1885. It included 432 varieties of apple and pear, and amongst them, at last, is the Beauty of Kent. Roger Deakin once described the story of the apple's

Plate VIII

5 Dymock Red

7 White Must

6 Munn's Red

3 The Red Foxwhelp

1 Rejuvenated Foxwhelp

2 The Bastard Foxwhelp

4 Black Foxwhelp

A plate from *The Herefordshire Pomona*, 1878–85, painted by Alice Ellis and Edith Bull.

evolution as 'as something between the Book of Genesis and the *Just So Stories*', and the Beauty of Kent is a late arrival on the stem of *Malus*, its unknown forebears having travelled thousands of miles before it begat the fruit which fell in front of Newton in Lincolnshire, and which had no recognised name for another century and a half. The *Pomona* records its first mention in a nurseryman's catalogue from about 1820, and illustrates it with a simple black-and-white cross section. But the description is lusciously and attentively sensuous, and you can glimpse the human fruitarian carrying on the work of those discriminating Chinese bears:

> Fruit, large, roundish, broad and flattened at the base, and narrowing towards the apex, where it is terminated by several prominent angles. Skin, deep yellow, slightly tinged with green, and marked with faint patches of red on the shaded side, but entirely covered with deep red, except where there are a few patches of deep yellow on the side next the sun. Eye, small and closed, with short segments, and set in a narrow and angular basin. Stalk, short, inserted, in a wider and deep cavity, which, with the base, is entirely covered with brown russet. Flesh, yellowish, tender, and juicy, with a pleasant subacid flavour.

Apple evolution continues. If the direction of modern mass cultivation and selling is entropic, favouring an ever-shrinking range of varieties and increasing uniformity of fruit, the apples themselves have different ideas. Cores left over from country picnics and thrown out of cars (and trains: Thoreau listed 'the Railroad Apple' in his Romantic lexicon) can sprout into wildings whose unpredictable fruit echoes the diversity of the fruit forests. One summer in the 1970s I was wandering along the beach at Aldeburgh in Suffolk when I chanced on a dwarf apple bush nestling between two shingle ridges. It came up no higher than my chest, and sprawled across maybe six yards of the beach. Goodness knows how much of it lay *under* the shingle. It had the beginnings of a few fruits, but I have never seen it at the right time to pick any. I discovered later that it's a well-known local curiosity, and the locals scrump the apples as soon as they have any semblance of ripeness.

No one knows how it arrived in this improbable position. It may have sprung from a thrown-out core or bird-sown pip that shingle movements buried serendipitously near an underground drift of soil or freshwater spring. Another theory is that it is the last relic of an eighteenth-century orchard, attached to a derelict cottage a hundred yards inland, which was buried as the coastline moved inexorably over it. This last tree stuck it out, not drowning but waving. Its topknot grows close to the stones, as densely twigged as a thorn bush.

However it originated, its endurance is remarkable. Pruned and pinched by the cold east winds, indifferent to dousings by salt spray at the highest tides, it must be one of the hardiest apple trees in Europe. I hope some enterprising fruitarian, mindful of the ever-rising sea levels, has taken grafts.

15

Intimations of Photosynthesis:
Mint and Cucumber

IN 1772 JOSEPH PRIESTLEY – preacher, dissenter and radical chemist – felt that he should inform 'my friends, and the public, that I have for the present, suspended my design of writing *the history and present state of all the branches of experimental philosophy*', because – an ageless author's complaint – he saw 'no prospect of being reasonably indemnified for so much labour and expence'. Such a titanic work by one of the brightest minds of the eighteenth century would have been something to behold. But what we got instead more than compensated: a study of the properties of 'different kinds of air' which included details of an exquisitely simple experiment that was to become one of the most important in the history of botany.

The nature of air obsessed Enlightenment scientists, including the group that Jenny Uglow called 'The Lunar Men', which included, in addition to Priestley, the poet Erasmus Darwin (Charles's grandfather) and the engineer James Watt. Many questions challenged their imaginations and experimental savvy. Which part of the air enabled animals and plants to breathe and grow, and which extinguished life? Did burning a substance release 'phlogiston' into the atmosphere, or remove some element from it? Priestley had been experimenting with ways of 'restoring' air which had been 'exhausted' by burning or breathing, and announces early in the substitute for his unwritten grand history (*Experiments and Observations on Different Kinds of Air*) that 'I have been so happy, as by accident to have hit upon a method of restoring air, which has been injured by the burning of candles, and to have discovered at least one of the restoratives which nature employs for this purpose. It is *vegetation*.' To be precise it was a sprig of mint. It is striking how *comfortable* were the whole lineage of pre-professional scientists – from Newton hunkered down in his farmhouse bedroom with a homemade rack of prisms, to Charles Darwin feeding carnivorous plants with the remains of his dinner – using the humdrum materials of domestic life in their experiments. I don't think this was entirely due to their being, willy-nilly, amateurs working at home. As the last non-specialists they must have also felt that what they were investigating was still existentially part of the warp and weft of their ordinary lives.

In the summer of 1771 Priestley had put a sprig of mint in a glass jar which he'd stood upside down in a bowl of water, presuming the plant would exhaust the air in the way that a breathing animal would. But months later it was still growing strongly, and he found that the air in the jar would neither extinguish a candle 'nor was it at all inconvenient to a mouse, which I put in it'. Over the weeks the root began to decay, and the mint put out ever smaller successions of leaves 'all the summer season'. He took care to extract any dead leaves in case they began to putrefy and affect the air. He also found that mint leaves would restore the air in a jar where a candle had burnt out. He tried different sorts of flame (wax, spirits of wine, brimstone matches), different plants (groundsel, cabbage) and pieces of plant. Spinach, splendidly, gave the best results, restoring a jar of 'burned air' in two days. But essential plant oils or bunches of leaves didn't work. 'The restoration of air,' he concluded, 'depended upon the *vegetating state* of the plant.' He begins to sense that plants affected the air in the opposite manner to animal respiration, and many hapless mice later, concludes that one mouse could live perfectly well in a jar of air made noxious by the breathing (and subsequent asphyxiation) of another of its kind simply by the insertion a sprig of mint for eight or nine days. After more timed and quantified trials later he was able to conclude that it

> cannot but render it highly probable, that the injury which is continu-
> ally done to the atmosphere by the respiration of such a number of

animals, and the putrefaction of such masses of both vegetable and animal matter, is, in part at least, repaired by the vegetable creation ... [and] if we consider the immense profusion of vegetables upon the face of the earth, growing in places, suited to their nature, and consequently at full liberty to exert all their powers, both inhaling and exhaling, it can hardly be thought, but that it may be a sufficient counterbalance.

The idea that sunlight, the motion of fluids through plants and their 'respiration' of air were in some way connected was not entirely new. Nor was the derision poured on these ideas by many outside science's inner circle. In 1727 Stephen Hales, rector of Farringdon in Hampshire (a parish adjacent to Gilbert White's Selborne), published the conclusions of his researches into 'vegetable statics'. Hales argued that the sun's heat caused water to be driven through a plant's sap vessels, and pass into the leaves to be 'perspired'. As so often in eighteenth-century science, thinking by analogy (in this case making a comparison with the way a warmed liquid expands and 'rises', as in a thermometer) grasped the overall effect, but not the exact mechanism. In fact water is transpired in trees by suction, as the change in pressure due to water loss from the leaves is transmitted down to the roots. In the 1735 edition of Jonathan Swift's surreal satire on the pretensions of contemporary science, *Gulliver's Travels*, Hale's experiments are the target of one of the funniest lampoons in the book. In Swift's Grand Academy of Lagado (a take-off of the Royal Society), one of the utopian scientists 'had been Eight Years upon a Project for extracting Sun-Beams out of Cucumbers, which were to be put into Vials hermetically sealed, and let out to warm the Air in raw inclement Summers'.

If Priestley had read Swift he was undeterred, and his own hermetically sealed vials proved to his satisfaction that the growth of green leaves (doubtless on cucumber plants as well as mint) was enabled by their exchange of common and burnt air, not yet named as oxygen and carbon dioxide. He'd also recognised a truth about the relation between vegetable and animal life, and established a fundamental principle of ecology. Benjamin Franklin, who had seen some of Priestley's flourishing plants during a visit from America, took Priestley's conclusion

The 'sunlight distillery' from *Gulliver's Travels*. An illustration by Milo
Winter from a 1912 edition.

one practical, and prophetic, step further: 'I hope this will give some check,' he wrote to Priestley, 'to the rage of destroying trees that grow near houses, which has accompanied our late improvements in gardening, from an opinion of their being unwholesome.' The President of the Royal Society sensed deeper global and egalitarian implications. 'In this the fragrant rose and deadly nightshade cooperate,' wrote Sir John Pringle in 1774; 'nor is the herbage, nor the woods that flourish in the most remote and unpeopled regions, unprofitable to us, nor we to them; considering how constantly the winds convey to them our vitiated air, for our relief, and their nourishment.'

Jan Ingenhousz quoted these words when he reported his own experiments in 1779, which showed that oxygen (it had been named by then) was given off only in sunlight, followed three years later by Jean Senebier's demonstration that the exhalation of oxygen in light depended on the intake of carbon dioxide. The basic principles of photosynthesis, the process that underpins all life on earth, had been established.

The central dynamic of photosynthesis – the conversion of sunlight into plant tissue via the fixing of atmospheric carbon dioxide – has been an inspiring motif for poets and writers ever since, the central idea being passed on from generation to generation, much as the molecules of life are. Erasmus Darwin, who kept an acute eye on developments in science, wove the discovery into his epic hymn to botanic creation, *The Economy of Vegetation*, in 1784. Erasmus felt obliged in his poetry (not, fortunately, in his prose) to animate botanical processes through the elaborate affairs of mythological humans. If you ignore the florid posturing, and the fact that he got the gas wrong, there is something energising in his thunderous and precocious image of a living earth:

> SYLPHS! From each sun-bright leaf, that twinkling shakes
> O'er Earth's green lap, or shoots amid her lakes,
> Your playful band with simpering lips invite
> And wed the enamoured OXYGEN to LIGHT ...

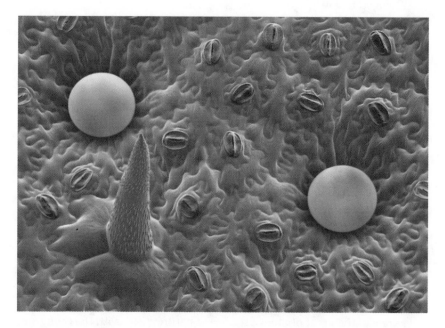

Close-up of a mint leaf taken through an electron scanning microscope. The blue spheres in the midst of the chlorophyll-rich, photosynthesising tissue are oil glands, which give mint its characteristic scent.

Whence in bright floods the VITAL AIR expands,
And with concentric spheres involves the lands;
Pervades the swarming seas, and heaving earths,
Where teeming Nature broods her myriad births;
Fills the fine lungs of all that *breathe* or *bud*,
Warms the new hearts, and yes the gushing blood;
With life's first spark inspires the organic frame,
And, as it wastes, renews the subtle flame.

Coleridge thought Erasmus 'a wonderfully entertaining and instructive old man', and his own take on photosynthesis echoes the pre-Gaian fervour in this stanza, though the science is both tougher and more metaphysical. The following passage, which I've condensed slightly, is buried in an appendix to Coleridge's densely theological tract *The Statesman's Manual* (1816) and hints at the key Romantic belief that the processes of the imagination resonated with those of organic growth itself:

... I seem to myself to behold in the quiet objects, on which I am gazing, more than an arbitrary illustration, more than a mere simile, the work of my own Fancy! I feel an awe, as if there were before my eyes the same Power, as that of the REASON – I feel it alike, whether I contemplate a single tree or a flower, or meditate on vegetation throughout the world, as one of the great organs of the life of nature. Lo! – with the rising sun it commences its outward life and enters into open communion with all the elements, at once assimilating them to itself and to each other. At the same moment it strikes its roots and unfolds its leaves, absorbs and respires, steams forth its cooling vapour and finer fragrance, and breathes a repairing spirit, at once the food and tone of the atmosphere, into the atmosphere and feeds *it*. Lo! – at the touch of light how it returns an air akin to light, and yet with the same pulse effectuates its own secret growth, still contracting to fix what expanding it had refined ... Thus finally, the vegetable creation, in the simplicity and uniformity of its *internal* structure symbolising the unity of nature, while it represents the omniformity of her delegated functions in its external variety and manifoldness, becomes the record and chronicle of her ministerial acts, and increases the vast unfolded volume of the earth with the hieroglyphics of her history.

Forty years on, the usually sentimental writer on gardens and botanical taste, Shirley Hibberd, is moved to a soaring passage of insight and rigour (but which never loses sight of Victorian values) while contemplating photosynthesis:

> The atom of charcoal which floated in the corrupt atmosphere of the old volcanic ages, was absorbed into the leaf of a fern when the valleys became green and luxuriant; and there … [t]hat same atom was consigned to the tomb when the waters submerged the jungled valleys. It had lain there thousands of years, and a month since was brought into the light again, in a block of coal. It shall be consumed to warm our dwelling … ascent into a curling wreath to revel in a mazy dance high up in the blue ether; shall reach earth again, and be entrapped in the embrace of a flower: shall live in the velvet beauty on the cheek of the apricot; shall pass into the human body … circulate in the delicate tissues of the brain; and aid, by entering some new combination, in enducing the thoughts which are now being uttered by the pen.

It's a passage which uncannily prefigures the journey of the carbon atom in Primo Levi's majestic *The Periodic Table*, written 120 years later. In this the atom also experiences many photosynthetic embodiments, including a bunch of grapes, a cedar tree (from which it's chewed out by a wood-boring insect) and the grass which eventually produces a glass of milk. The atom in the milk is drunk by the writer and – just as in Hibberd's narrative – takes part in the neural synapse which guides his pen: 'a double snap, up and down between two layers of energy, guides this hand of mine to impress on the paper this dot, here, this one'.

Unravelling the process of photosynthesis in plants was arguably the most important development in the history of biology. Most forms of life on earth depend on this transformation of the sun's energy into living tissue. As the ethnobotanist Tim Plowman remarked, contemplating twenty-first-century revelations about plant communication, 'Why should that impress us? They can eat light, isn't that enough?'

Will we be more or less impressed when we are technologically able to replicate photosynthesis, to restore air or 'eat light' ourselves?

16

The Challenge of Carnivorous Plants: The Tipitiwitchet

THE NEW SPIRIT OF EXPERIMENTATION flourished in parlour laboratories and garden rooms, and found its way into the gossipy correspondence of colonial naturalists stranded continents apart. Arthur Dobbs, governor of New Carolina, was an adventurous gardener and naturalist who in 1750 had been the first person to publish a detailed account of the role of bees in pollinating flowers. One of his close friends and correspondents was the Quaker naturalist Peter Collinson, owner of a famous garden in Peckham, then a country village south-east of London. On 2 April 1759 Dobbs wrote to his friend with news of his own garden and the fortunes of the seeds they exchanged. It's a chatty but excited letter, catching the feeling that the new frontiers of Empire were in knowledge as well as territory. And he adds, almost as an afterthought, a note on a local 'sensitive' plant he had come across in the nearby swamps:

> I thank you for the cedar cones & almonds you sent me but as they were above eight months before I Received them none of them came forward ... as I now Live near the sea coasts [at Brunswick] and have taken a little Plantation at the sound on the sea coast, I intend to try oranges and Lemmones, as the Palmetto Cabbage Trees thrive there, and by preserving shelters from our northwestwards can make any experi[m] ents, and want to try Dates ... We have a kind of Catchfly sensitive which closes upon any thing that touches it[.] [I]t grows in the Latitude 34 but not in 35 degrees – I will try to save the seed here.

This letter, quoted in Charles Nelson's 'biography' of the Venus flytrap, is the first written reference to a plant whose extraordinary behaviour would for the next hundred years unsettle received ideas about the distinctive character of plants and their place in the natural scheme of things, and eventually suggest a possible candidate for the 'vital force' that Romantic thinkers believed gave life to plants. Since classical times, philosophical notions of the ordering of Creation had been dominated by the concept of the Great Chain of Being, a rigid and unbreakable hierarchy, with God at the top and inanimate rocks at the bottom. In the biological part of the spectrum animals were higher than plants, and possessed of superior powers and abilities. A 'sensitive' plant, with powers of movement and capture conventionally believed to be the prerogatives of 'higher' organisms, undermined the whole concept of the Chain – at least in the eyes of theologians and more traditional scientists. But the discovery of the flytrap occurred at a propitious moment, when poets and the new generation of romantically inclined scientists were beginning to question the validity of the hierarchical model, and ask whether the distinctions between animal and plant were as rigid as had always been assumed. The debate sparked off by the flytrap had epistemological repercussions too. It put the usefulness of biological analogy – a favourite eighteenth-century mode of 'explanation' – to severe test. The common assumption that superficial similarities in behaviour implied similar causative processes had led to a multitude of blind alleys and astigmatic blunders. But often the intelligent use of analogy opened the way to new understandings of the way that plants lived out their lives, and this is the way it was, in the long term, with the flytrap. Flesh and grass, after all, were ultimately governed by the same physical laws.

Collinson, his appetite whetted by Dobb's offhand announcement about the 'Catchfly sensitive', was eager for more information, and in January 1760 Dobbs obliged him. His description is made especially vivid by its switching between simile and closely observed detail, but it is nonetheless a classic example of the mechanistic model of a plant.

> [This] great wonder of the vegetable kingdom is a very curious unknown species of sensitive; it is a dwarf plant; the leaves are like a narrow

segment of sphere, consisting of two parts, like the cap of a spring purse, the concave part outwards, each of which falls back with indented edges (like an iron spring fox trap); upon any thing touching the leaves or falling between them, they instantly close like a spring trap, and confine any insect or any thing that falls between them; it bears a white flower: to this surprising plant I have given the name of Fly Trap Sensitive.

These days, this carnivorous oddity is known almost universally as the Venus flytrap, and a rare endemic of the Carolina swamps has become a mail-order novelty for conservatories and windowsill displays throughout the world. Schoolboys relish tweezering flies (dead and alive) into its spiny maws and do not find the existence of a plant which eats flesh the least bit peculiar. That it was seen as an altogether more challenging organism in the eighteenth century – intellectually exciting or morally suspect, depending on your point of view – is evident in the hunger that grew for specimens and seeds, though very few of them reached their destination alive.

In 1762 William Bartram, son of another Quaker naturalist, John Bartram of Philadelphia, paid Dobbs a visit and took some living flytraps back to his father. Shortly afterwards, John Bartram wrote to Collinson describing 'sensitive catch-fly' as a 'Wagish' plant, and relayed that it had the name of 'tipitiwitchet' amongst the local Native Americans. Collinson sunk further into frustration. A 'Wagish' tipitiwitchet sounded as if it might be a much more exotic visitor than the packets of obstinately sterile seeds and desiccated specimens he'd so far received.

Nothing much happened on the flytrap front for another four years, until William Young, an ambitious Philadelphian nurseryman, turned up in London. Young had built up an influential English clientele, toadied to the royal family by making spontaneous gifts of seed, and was eventually invited to Court for a year, where he described himself as 'Botonist to their Majesty's' (sic). But he disgraced himself by being 'drawn into very bad Company of women', and squandered his earnings. Queen Charlotte settled his debts on condition that Young returned to America immediately, and stayed away from Kew, permanently. So during 1767 he toured

Venus flytrap, the tipitiwitchet, painted in 1847, with the vulval pink surfaces of the leaf-traps rendered very realistically.

the Carolinas and collected 'many barrels filled with plants', including a few tipitiwitchets. Then his ambition overcame his loyal subservience and, flouting the queen's edict, he decided to take some specimens back to England. They were packed in damp moss and survived the journey. Young sold them to James Gordon, a nurseryman friend of Collinson, and by the summer of 1768 the first live tipitiwitchets to have reached England were growing well in a London glasshouse.

At the end of August one of them came into flower, an event which caught the attention of John Ellis, another of Collinson's friends. Ellis was a typical example of the Enlightenment's polymorphous and influentially networked impresarios. He was a Fellow of the Royal Society, hobnobbed by post with Benjamin Franklin and Linnaeus and had a day job as a linen merchant alongside his consuming amateur passions for natural history and gardening. Straight away he commissioned a series of professional drawings of the plant, and took it upon himself to complete a full Linnaean name. The botanist Daniel Solander had already named its genus (entirely new to science) as *Dionaea* – 'from the beautiful Appearance of its Milk-white flowers, and the Elegance of its leaves, [I] thought it well deserved one of the Names of the Goddess of Beauty'. The specific name Ellis chose was *muscipula* (meaning mousetrap, oddly, not flytrap), and he published his account – perhaps with an eye to maximum popular appeal – in a London newspaper, *The St James Chronicle*. The *Chronicle* may not have had the kudos of the *Transactions of the Royal Society*, but first publication is authoritative, and on 1 September 1768 the 'Catchfly sensitive' and the 'Wagish tipitiwitchet' were formally dubbed *Dionaea muscipula*, Venus's flytrap (or as Charles Nelson literally translates it, 'Aphrodites's mousetrap'). With the hindsight of 250 years, there is already the slightest whiff that all was not as it seemed in the naming of the siren plant.

The name's curious, almost oxymoronic linking of the Goddess of Love with a spiked gin trap tickled public imagination even further, and letters, pamphlets and engravings flew about botanical circles. Later in September 1768, Ellis wrote to Dr David Skene, describing graphically the apparatus inside the leaves:

In the expanded leaf are many minute dots in both lobes: these magnified are of the figure of a compres'd raspberry or arbutus fruit, and of a fine red colour. In these lie the sensibility of this Plant. I have called them the irritable glands of this plant; But nature has still further views in well securing the Insect, than by the lobes coming together, and the rows of Spines clasping each other. There are amid the glands on each lobe three erect spiculae, which must either run into the Insect on the closing of the lobes, or at least prevent it wriggling to and fro to effect its escape.

He then adds a note which took the puzzle of the organic order to which the flytrap belonged into hazier and more troubling territory:

Lord Moreton asked me a shrewd Question, when I shewd the drawing at the Society the day of K[ing] of Denmarks Election. Do you think Sr that the plant receives any nourishment from Insects it catches? I acknowledg'd my ignorance.

Catching and caging flies, even running them through with lances was one thing; they could quite conceivably be purposeless reflexes, or some sort of self-protection by the flytrap. But *digestion* of its hapless prey was an altogether more dyspeptic idea, even to the adventurous imaginations of botany's inner circle.

What is odd about Ellis's otherwise excellent description is the confidence with which he explains away the flytrap's behaviour as an example of 'irritability'. This was a popular concept amongst reductionist and analogical botanists, used to account for any otherwise inexplicable excitability in plants, as if it were an a priori property of vegetable tissue, as it is of human skin. A young Edinburgh student, William Logan Jr, who had heard Ellis lecture, wrote to him with an equally vague idea, which seemed both a credulous throwback to the world of medieval bestiaries and the Great Chain, and a glimpse into the future of biology:

Excuse me if in vanity & fullness of my heart I now enter on my own opinion viz: that there is a chain by which plants and animals are connected and that there is an Amphibious State neither entire Plant nor Animal. We have heard of Mermen & Mermaids. We have seen Sea

Dogs & Sea Lyons. We have the Bat and numerous Instances in other parts of the Creation where the animal belongs to two classes. Why not then in Plants?

Intellectuals were intrigued by the existential status of the plant, and its technique in catching prey, but not many asked questions about the purpose of the fly catching. Most seemed happy to accept the model of a mechanical spring trap driven by some otherwise unexplained 'irritability', including Charles Darwin's ever ingenious grandfather, Erasmus. He had seen the 'Flytrap of Venus' for himself on 20 August 1788 at Ashburn Hall in Derbyshire, and had drawn 'a straw along the middle of the rib of the leaves as they lay upon the ground round the stem' to watch them close 'in about a second of time … like the teeth of a spring [rat]-trap'. Erasmus thought the plant's aggressive behaviour was to protect the leaves from nibbling insects, not eat the marauders themselves. This was also, he believed, the reason behind the fly-catching habits of *Dionaea*'s European relative, the sundew, which holds and smothers insects by means of sticky hairs on its leaves, 'As the ear-wax in animals seems to be in part designed to prevent fleas and other insects from getting into their ears.'

This is a footnote to a stanza on sundew ('A zone of diamonds trembles round her brows') in Erasmus Darwin's preposterous but hugely entertaining *The Botanic Garden*, a long, florid poem – subtitled *The Loves of Plants* – which attempts to translate Linnaeus's sexual system for classifying plants into a series of melodramatic tableaux, full of sylphs, swains, knights, wronged maidens and vengeful demi-gods. When he gets to *Dionaea* (which he also classes as a *Silene*) he has nothing to say about its life, but his brief couplet has an odd and probably unintentionally erotic resonance with what are now thought to be the origins of that teasing name 'tipitiwitchet':

The fell SILENE and her sisters fair
Skill'd in destruction, spread the viscous snare.

190

In December 1990 Daniel L. McKinley, an emeritus professor of biology from New York, wrote to Charles Nelson about some tipitiwitchet evidence he'd unearthed while working on a biography of William Bartram. He'd been puzzled, as have many naturalists, over this supposedly vernacular tag, and the vague suggestions by Ellis that John Bartram had reported it as an 'Indian name, either Cherokee or Catabaw [sic] but I cannot now recollect'. What alerted McKinley's suspicions was a throwaway remark in a letter from Collinson in June 1784: 'I hear my Friend Dobbs at 73 has gott a Colts Tooth in his head & married a young lady of 22. It is now in vain to write to him for seeds or plants of Tipitiwitchet now He has got one of his Own to play with.' McKinley was told by authorities from the Department of Anthropology at the US National Museum of Natural History that no such word existed in any native American language. (Since then I've located, via an internet search, a very similar term for the flytrap – 'titipiwitshik' – in a dictionary of the Lenape language of East Coast Indians. This is certainly close enough to be a prompt for the snappier and more memorable term the North Atlantic botanists' club began to use, but it doesn't invalidate McKinley's intriguing theory about why they may have tweaked it.)

McKinley then tried another tack, and looked for what the word might mean in English. In Eric Partridge's *Dictionary of Slang and Unconventional English* (1970), 'tippett' is a fur collar. This is one of the standard meanings given in the *Oxford English Dictionary*, too, which also gives as meaning (2), 'a jocular term for a hangman's noose', and (3), 'an organ or feature in an animal resembling or suggesting a tippet'. In other dialect dictionaries McKinley found 'twitch' as a noose for frisky horses, or a tight boot. 'Twitchety' is fidgety or jerky. Ozark mountain folk tales used the term 'twitchet' for the female genitalia. It begins to look as if flytrap leaves – a pair of moist, red, semicircles fringed by hairs, which remorselessly gripped their hapless prey – were being likened to the *vagina dentata*. Whichever member of the group originally coined the name, it looks as if he was deliberately intending it to be a piece of covert ribaldry.

McKinley also suggests that the 'Venus' in Venus flytrap was a respect-ably disguised riff on the same idea. In Botticelli's painting *The Birth of*

Venus, the Goddess of Love emerges – naked but demure – from a scallop, a rather flat and unsuggestive mollusc (the painting is often jokingly referred to as 'Venus on the half-shell'). But *Venus* as a scientific name was proposed by Linnaeus in 1758 for a group of bivalves, including the Quahog shore clams of North America (now generically known as *Mercenaria*) and the Royal Comb Venus shell (*Venus dione* now *Pitar dione*), which when opened have an uncanny resemblance to the splayed leaves of a tipitiwitchet. They're moist, semicircular, full of soft, palpable flesh, and with a powerful grip when closing. A man trap more than a *muscipula*.

This is all conjecture, but highly plausible, and in the end not at all surprising. This group of middle-class literati were simply adding their ha'p'orth to the long vernacular tradition of appointing bawdy names to plants, especially when there is the slightest hint of a sex organ in their appearance. (Others include bull's bag for orchid roots, dog's cock and jack-in-a-box for cuckoo pint.)

It's gratifying to see this tradition still continuing, with witty contemporary grace notes. In 2006 the carnivorous plant breeder Barry A. Rice decided to dub a well-known but till that point unnamed bright green flytrap, 'Justina Davis'. William Collinson had been over discreet in describing the seventy-three-year-old Arthur Dobbs's bride as being twenty-two-years-old. Justina was just fifteen. Rice explained his tag: 'The electric green leaves can be mistaken for fresh young traps that have not yet time to develop pigmentation.' In 1968, secondary school children on the Isle of Man, acting from a kind of instinctive sympathetic magic, also saw 'fresh young traps' in the ensnaring leaves of sundew. In the year of peace and love, a labelled dish of the plant in the Manx Museum was pilfered by local teenagers, who slipped bits of the plant into the pockets of their crushes.

Seventy years after Erasmus Darwin had tickled the maw of Venus with a straw, and seen her snap tight shut on it, his grandson Charles became intrigued by the more decorous entrapments of *Dionaea*'s relative the sundew:

During the summer of 1860, I was surprised by finding how large a number of insects were caught by the leaves of the common sundew (*Drosera rotundifolia*) on a heath in Sussex. I had heard that insects were thus caught, but knew nothing further on the subject. I gathered by chance a dozen plants, bearing fifty-six fully expanded leaves, and on thirty-one of these dead insects or the remnants of them adhered … Many plants cause the death of insects (for instance the sticky buds of the horse chestnut (*Aesculus hippocastanum*)) without thereby receiving, as far as we can perceive, any advantage; but it was soon evident that *Drosera* was excellently adapted for the special purpose of catching insects, so that the subject seemed well worthy of investigation. There results have proved highly remarkable.

Charles Darwin was as ingenious a writer as he was a scientist. In this passage he sets up his forensic investigations with the craft skills of a country-house thriller author: the drowsy backcloth of a Sussex heath in summer, the mysterious bodies, the confession of initial ignorance, the promise of startling evidence to come. *Insectivorous Plants*, which progresses to the mystery of the even more murderous flytrap, is a page turner.

Sundew haunts bogs and damp acid moorland, where the soil is low in nutrients. Its most significant feature is its leaves, which, as grandfather Erasmus had noted, are round and covered on their upper surface with fine, translucent, dew-tipped hairs – 'or tentacles as I shall call them,' Charles writes, 'from their manner of acting'. It is the first of many animal analogies he uses in his deconstruction of botanical carnivory. He counts the numbers of tentacles on thirty-one leaves, and finds that they range from 130 to 260. Each is 'surrounded by large drops of extremely viscid secretion, which, glittering in the sun, have given rise to the plant's poetical name'. They are short in the centre of the leaf, and longer towards the edge.

The problem they posed Darwin is this. When a small object is placed on the central hairs, they appear to transmit 'a motor impulse' to the marginal hairs, which bend slowly towards the centre until they all become closely curled around the object. Full closure takes between one and five hours, depending on the size of the object and its nature.

Fly trapped in the viscous spines of sundew.

How did vegetable tissue move purposefully to catch insects, and then go on to digest them? What set the whole sequence in motion? How was the 'motor impulse' transmitted through the leaf?

Over the next months (possibly years) Darwin conducted an astonishing range of experiments with sundew plants, to test their sensitivity and appetite. He tries feeding them insects, dead and alive (but is soft hearted enough to rescue a half-caught gnat), and then with much of the contents of his larder: hard-boiled egg, roast beef, milk, cheese. Then he tries splinters of glass and cinders, and a huge range of chemicals. He breathes on the leaves, and even coughs up some of his own sputum to test the limits of the sundew's palate for organic material. He uses minute and precisely measured lengths of women's hair, 'weighed for me by Mr Trenham Reeks, in an excellent balance, in the laboratory, in Jermyn Street'.

His findings were exhaustive and beyond dispute. A living insect was more efficient as a trigger than a dead one, as in struggling it presses against more glands (Darwin's name for the viscous tips of the tentacles). A gland will be excited by being simply touched three or four times, and after a while the leaf itself begins to curl inwards 'so as to form a temporary stomach'. The leaf quickly distinguishes between organic and inorganic matter (implying some sort of chemical feedback from the digestive system which Darwin doesn't explore), but when satisfied that its catch contains nitrogenous matter starts to pour out digestive juices. Darwin makes the point that these act in the same manner as human gastric juices, and though unable to analyse the secretions, found that they became more acidic during digestion, and contained some analogue of pepsin in their capacity to dissolve albumen. 'There is therefore,' he concludes, 'a remarkable parallelism between the glands of *Drosera* and those of the stomach in the secretion of their proper acid and ferment.' In analysing the digestive process he also established the link between *Drosera*'s nutrient-poor habitats and its insect-catching habit: the 'advantage' that he was unable to see at the start of his investigation was a compensation for the lack of nutrients in acidic, peaty soils. (There is an endearing aside where the Victorian dietary improver gains ascendancy over the stringent experimenter: 'a decoction of cabbage leaves is far more exciting and probably nutritious to *Drosera* than an infusion made with tepid water'.)

He repeats a similar series of experiments with the Venus flytrap, with results that were comparable, but strikingly different in detail. The lobes of the flytrap's leaves are covered with minute glands which have the ability to secrete digestive juices and absorb food. Each blade of the leaf also has three prominent filaments which are highly sensitive to touch. Darwin observed that they were selective in their response (they ignored simulated raindrops and human hairs) but didn't realise that at least two of the filaments had to be touched within less than twenty seconds for the trap to be sprung. (A mechanism to ensure the plant doesn't waste its energies catching the wind or insignificant prey.) When this happens the lobes snap shut in about one tenth of a second and begin squeezing the insect with such force that an impression of it can be imprinted in the lobe's surface. The glands start to pour out digestive juices. Sometimes, when the prey is large, they will stay rigidly shut for up to ten days, and can't be prised open without tearing the leaf tissue.

Darwin himself didn't uncover the channel or mechanism by which the stimulus from the filaments activated movement of lobes or tentacles. He observed that there was a movement of fluids, and that waves of 'aggregation' moved through the activated plant. But he knew these were the effects rather than the source of the excitation that spread through the plant. He was thrilled when in 1874 one of his contemporaries, John Burdon-Sanderson, discovered that an electric potential was set up across the leaf when one of the hairs is touched. Burdon-Sanderson was a professor of physiology at University College London, and had made a study of the electrical impulses which activated animal muscles. And again the fastidious Darwin was happy to make a cross-kingdom analogy when reporting the professor's findings, likening the lobe's movement to the 'contraction of the muscle of an animal'. It was to be more than a century later that the mechanism of the two-hair trigger was unravelled. Two researchers at the University of Bonn, followed by Alexander Volkov at Oakwood University in Alabama, found that the first touch of a single hair activated an electric potential, which is stored, as if in a temporary battery, by a rise in the concentration of calcium ions in the lobe. This lasts about twenty seconds and then dissipates. But if by then a second hair has been touched, another potential builds up, the cumulative charge overrides some cellular fail safe, sets off a shift of fluids in the leaf cells, and the trap snaps shut. In effect *Dionaea* has a short-term memory, storing information about its earlier experiences and using this to influence its future behaviour.

Not only had electric signalling been discovered in a plant, fulfilling the dreams of many eighteenth-century botanical thinkers, but also the capacity to keep and retrieve information – an attribute previously thought to be the prerogative of brains. Erasmus Darwin would have been beside himself.

17

Wordsworth's Daffodils

THE DAFFODIL has been at the centre of many botanical performances – extended allegories, seasonal pageants, bizarre analogies, comedies of mistaken identity. Its name is probably an anglicised version of asphodel, a forgivable attempt to claim one of our own wild lilies as the flower of the Elysian Fields. According to the subtitle of one modern guide it is 'The most popular spring flower in the world', and if a garden has to be reduced to a single flowerpot or vase, you can be fairly sure what its occupants will be in February and March: a fanfare of golden trumpets on a wand – brassy, forthright, sun-mimicking, spring-conjuring, the first real colour of the year.

Daffodils have also been deeply involved in the long debates about the use of plants as metaphors for human feelings, and indeed about whether plants have feelings and vitality of their own. Wordsworth's 'I wandered lonely ...' – the best known flower poem in English, but with no ascribed title – is, depending on your point of view, a joyous song to the union of humans and nature, or a prime example of where the Romantic sensibility descended into sentiment, granting human emotions to clouds and country dance steps to flowers.

> I wandered lonely as a cloud
> That floats on high o'er vales and hills,
> When all at once I saw a crowd,
> A host of golden daffodils;
> Beside the lake, beneath the trees,
> Fluttering and dancing in the breeze.

Crediting plants with the ability to 'dance' rather than, say, 'wave' is, many would argue, the classic example of a fault line that has run for millennia through our attempts to describe the natural world. Instead of notating it on its own terms (whatever that might entail) or at least in neutral language, we use analogies drawn from human behaviour. Wordsworth *embodies* the daffodils, and with that absorption into a human frame conjures up the possibility of all the affective and behavioural business that accompanies human action – intention, feeling, narratives with meanings. Worse, in this particular instance, Wordsworth is guilty of what John Ruskin would later call 'the pathetic fallacy', the grafting of his own emotional state onto the unconscious workings of nature: clouds in the Lake District are, on the whole, very rarely lonely. The Romantics' response to this objection was to turn it on its head, and insist that they were not so much forcing nature into a human shape as seeing the natural expressing itself in the human, body and spirit.

The daffodil may be a cipher in this argument but it also has a story of its own that helps in understanding the Romantics' view of the life of plants. Regardless of its status as a metaphor in Wordsworth's poem, the daffodil has long been a metaphor *for* metaphor, for our human compulsion to see ourselves reflected in the natural world, a habit which goes back as far as the Roman poet Ovid. He wrote his famous *Metamorphoses* early in the first century AD as a series of moral fables in verse, in which inappropriate behaviour amongst the gods and their associates is punished or resolved by shape shifting. Ovid's version of the story of Narcissus is based on an ancient Greek myth about a preening boy who is turned into a flower. Narcissus is a strikingly pretty, androgynous teenager, and is well aware of his good looks. He is pursued by the nymph Echo and throngs of adoring naiads and oreads, but pride and self-obsession make him reject them all. The goddess Nemesis steps in to take revenge on the nymphs' behalf. Narcissus is led to a silver pool, and dropping down to drink, sees an image in the water, and is transfixed with love for it. Ovid's description is as precise as a field guide:

> Smooth cheeks, and ivory neck, and the bright beauty
> Of countenance, and a flush of colour rising
> In the fair whiteness. Everything attracts him
> That makes him attractive.

Narcissus tries to kiss and embrace the figure but it dissolves at his touch, and at last he twigs that it is his own reflection. 'He is myself! I feel it! I know my image now. I burn with love/ Of my own self!' The mirrored surface of the water displays then separates him from the object of his ardour. He collapses in despair, dies and withers away, leaving only 'a flower with a yellow centre/ Surrounded by white petals.'

Most early versions of the myth conclude with Narcissus committing suicide, and Ovid is the first storyteller to flourish, like a conjuror with a hat, a real flower as the climax of his performance. It wasn't called 'Narcissus' in Roman Italy, and doubtless had many regional tags (in modern Italian it is *fior-maggi* – literally, 'top flower'), but is easily recognisable as the species, common in the meadows of southern Europe and in gardens everywhere, which is known in English as poet's, or pheasant's-eye, narcissus. Carl Linnaeus gave it the species suffix *poeticus* in the mid eighteenth century. But the genus name *Narcissus* dates back at least to the early sixteenth century, when classically literate botanists began the business of ordering and naming species. Ovid's story and exact description were too good to leave out of the grand narrative they were beginning to create. We can imagine them admiring his fastidious poetic method as he homes in on this species to make his point – its solitary elegance, its aloofness, the intensely focused face, the 'flush of colour rising in the fair whiteness of the petals'. Yet when the French painter Claude Lorrain produced one of the best known visual renderings of the story in 1644, *Landscape with Narcissus and Echo*, he includes a quite different species of daffodil. The myth's other ingredients are reproduced faithfully. In the centre of the painting a desperate looking Narcissus crouches in front of a forest pool, mesmerised by his own reflection. Echo poses naked but neglected on a rock; nymphs, perched up trees like cooing doves, gaze down at Narcissus adoringly. And just visible in the right foreground, next to a formal clump of greying burdock leaves, is a tuft of the wild daffodil, *Narcissus pseudonarcissus* (the suffix means 'false narcissus' in a botanical sense, but the namers would have been aware of their cultural pun). It is pert and shining yellow. In modern Italian its vernacular name is *trombone*. Another view, perhaps, of Narcissus's histrionic character.

The daffodils had come a long way by the seventeenth century. The Elizabethan botanical writer John Gerard lists thirty-five species and varieties already in cultivation by 1597, including the splendidly dubbed '*Narcissus media croceus seroinus Polyanthos*, the late many floured Daffodil with the Saffron-coloured middle'. The first written record of *N. pseudonarcissus* is by William Turner, half a century before. Gerard's account of this species – or rather his absence of an account – suggests a flower as widespread and well known as the bluebell: 'the common yellow Daffodill or Daffodowndilly is so well known to all that it needeth no description ... [It] groweth almost everie where throughout England.' And it had vernacular names throughout England too, suggesting an ancient familiarity: Butter and Eggs in Somerset, Churn in Lancashire, Daffydilly in Northamptonshire, Giggary in Devon, Julians in Hertfordshire, Easter Lily throughout the West Country and Lent Lily just about everywhere. When the Belgian botanist Charles de l'Écluse visited London in 1581, he'd seen country women selling bunches of wild dillys across the City, and that 'all the taverns may be seen decked out with this flower'. It's a cheering thought: Cockney pubs done up like parish churches at Eastertide.

Two centuries later this was the species that William Wordsworth immortalised in his poem from 1804. In it the daffodils – 'Fluttering and dancing in the breeze', 'Continuous as the stars that shine' – are active agents, excitable jigglers, flowers changed into humans. But in the key last stanza Wordsworth does something analogous to Ovid, and changes a human (himself) into a kind of flower – except that this is not some mythic transfiguration, but a conjuring of real-world resonance:

> For oft, when on my couch I lie
> In vacant or in pensive mood,
> They flash upon that inward eye

> Which is the bliss of solitude;
> And then my heart and with pleasure fills,
> And dances with the daffodils.

Dances *with* the daffodils, not *as* one, nor as their dancing master. I imagine Wordsworth believed the flowers were complicit in the dance, and not simply mechanical windmills. In 'Lines Written in Early Spring' (1798, when he was twenty-eight) he'd created another windswept portrait of this central Romantic belief, that nature and the human spirit moved to the same rhythm:

> And 'tis my faith that every flower
> Enjoys the air it breathes ...
> The budding twigs spread out their fan,
> To catch the breezy air;
> And I must think, do all I can,
> That there was pleasure there.

I don't think that this, or the sprightliness so wonderfully captured in the daffodil poem, implies that Wordsworth saw plants as possessing consciousness. But he thought that they possessed an 'active principle' beyond their surface characteristics, with an agenda of goals and rewards. The Romantic poets – and many of their scientific contemporaries – had a hard time from twentieth-century critics for perpetuating the idea of analogies between plants and animals. But the most recent botanical discoveries are showing that their intuitions about plant sensitivity, however bizarrely arrived at and argued, weren't fantasies.

When I first became ensnared by plants in my twenties I imagined I was a Romantic, but it was Gerard's witty, down-home banter that warmed me more than Wordsworth's ecstatic wordiness. I loved the way 'old John' had largely avoided appropriating plants as mute servants in the great Enlightenment project of mastering nature, and instead had stuffed them affectionately into his memory and garden corners, and probably into his pockets and hair too. They were his Significant Objects. In his *Herball,*

or General Historie of Plantes, sixteenth-century London is pictured as a meadow which happened to grow buildings too. Navelwort sprouted in Westminster Abbey over the door that led from Chaucer's tomb to the Old Palace. Its fleshy leaves were good for feet made sore pounding the City's streets, which could be soothed by 'being bathed therewith, or one or more of the leaves laid upon the heele'. Musk mallow, in different moods, ornamented both 'the left-hand side of the place of execution called Tyborne' and 'the bushes and hedges as you go from London to a bathing place called Old Foorde'. And in a wood at Hampstead, Gerard – discoverer as well as surveyor – made what is the first written record for the common spotted orchid ('Female Satyrion Royall [with] … a tuft of purple floures, in fashion like unto a Friers hood'). He pictures the whole of Britain like one of Giuseppe Arcimboldo's portraits in vegetables, a society jigsawed together out of plants that played tricks, cured sore feet and broken marriages, ornamented the world as colourfully as costumes, marked the march of days and the contours of the land. Gerard made them into neighbours, if not exactly friends.

By contrast, I regarded Wordsworth's plant writing – I didn't know a great deal beyond the daffodil verse – as sentimental and unobservant. I couldn't take those vaporous clichés about the solitude of clouds, and blooms like stars that 'twinkle on the milky way'. The poet's gaze I felt (quite sternly) should have been directed towards the green earth, not up to the heavens. I also felt in some vague and snobbish way that his sentimentality was partly responsible for the suburbanisation of English country lanes by blowsy nurseryman's cultivars. I hadn't understood the poem at all. More to the point, I had never seen the wild species which inspired it. I knew its elegance from pictures – sharp, bicoloured, more art deco than Laura Ashley – but had never seen it out in the wind, beneath the sky.

But I recognised that William had done something special in his image of the golden host 'fluttering and dancing in the breeze'. He'd *animated* a flower, given movement to a still life. Ted Hughes, in his superb poem about gathering wild daffodils to sell when he was a child, puns them as 'windfalls'. Innocent of Giggaries, I had to take their tremulous energy on trust. But I had read about, and been intrigued by, another kind of daffodil movement – an enforced retreat, a stubborn digging-in. The species had declined dramatically since Gerard's time, but Geoffrey Grigson (his *The Englishman's Flora* was my Baedeker and Bible in the 1970s) had said that the best place to see them was in Gloucestershire. Around Newent and the hills towards Herefordshire, they still flourished in the woods and meadows. I found a poem by Lascelles Abercrombie, who came to live at Ryton, near Newent, in 1910, which mapped this blessedly endowed region with loving exactness:

> From Marcle Way,
> From Dymock, Kempley, Newent, Bromesberrow,
> Redmarley, all the meadowland daffodils seem
> Running in golden tides to Ryton Firs ...

This is a map of half-hidden treasure, the village names chiming the steps you take to find the golden tides. In the mood of pastoral nostalgia that suffused the troubled 1930s, this tract of country became known as 'The Golden Triangle', and the Great Western Railway laid on 'Daffodil Specials' for weekend tourists who came to walk the yellow road and buy bunches at farm gates. I imagined it as a floral refuge, an arcadia which had escaped the agricultural ravages that had driven the flower from so much of the countryside.

It wasn't till my early thirties that I got to see the Golden Triangle for myself, and have my own daffodil epiphany. I'd had a sickly winter, cooped up indoors, and in late March an understanding friend and fellow botanophile offered to drive me to west Gloucestershire to see the show. As soon as we had left the motorway the daffodil carpet unrolled exactly as in Abercrombie's verse, but with a ragamuffin promiscuity that I hadn't expected. We'd driven in along the M5, and no sooner had we turned west onto the Ross Spur than they were tumbling in rivulets down the embankments. We saw our first crowded round an electricity substation. Further along we found them studding the brown stands of bracken in a forestry plantation ride. Once we'd left the main roads and were meandering along the network of lanes between the villages – Marcle, Bromesberrow, Redmarley – they were so profuse and peculiarly sited it felt like a tease: 'guess where we'll pop up next'. They lined the

Walter Crane, *La Primavera* (detail), 1883.

edges of paddocks and ditches, clumped round gravestones and at the feet of road signs, filled orchards and copses, and even grew through the blades of winter wheat in the arable fields. At Kempley (where the churchyard had islands of daffs neatly spared in the mown grass) we talked to the vicar, who explained how farmers and fruit growers had regarded them as a wild catch-crop, and how, paradoxically, the picking and selling for the tourist trade had helped ensure they were valued and looked after. He suspected they had also been deliberately spread about a bit. Now they were the region's totemic flower, with nature trails and daffodil walks marked out and running through 'all the meadowland ...'

Wordsworth's 'daffodil experience' had also been a social occasion. On 15 April 1802 he and his sister Dorothy had been to visit their friends the Clarksons in Eusemere near Ullswater. They were in a buoyant mood, together again after a long interval, and William was excited about a visit to his fiancée Mary Hutchinson to arrange details of their forthcoming marriage. They set out after dinner to walk the twenty miles back to Grasmere – a mere stroll for these long-distance trampers. There was a gale blowing, and Dorothy records in her journal how they sheltered under a furze bush: 'The wind seized our breath.' They followed the western edge of Ullswater, where there were 'a few primroses by the roadside – woodsorrel flower, the anemone, scentless violets, strawberries, and that starry yellow flower which Mrs C. calls pile wort'. Then, just as they were passing the woods beyond Gowbarrow Park, they spotted the daffodils. Just a few at first, close to the waterside. Dorothy writes:

> We fancied that the lake had floated the seeds ashore, and that the little colony had so sprung up. But as we went along there were more and yet more; and, at last, under the boughs of the trees, we saw that there was a long belt of them along the shore, about the breadth of a country turnpike road. I never saw daffodils so beautiful. They grew amongst the mossy stones about and about them: some rested their heads upon these stones as on a pillow for weariness; and the rest tossed and reeled

and danced, and seemed as if they verily laughed with the wind that blew them over the lake.

Dorothy's bubbling excitement at being with her brother again is unmistakable in those inclusive we's, and in her fancy that the lake had, as it were, floated the seeds to their feet, to prepare a flower-strewn pathway. It's not inconceivable that this is how they arrived there, though wild daffodils were doubtless anciently established throughout Cumbria. The first *Flora* of the county records them as being in the Lakes in 1690. The Gowbarrow colony continues to flourish today and still grows amongst mossy stones exactly as Dorothy describes. But as is evident from Gloucestershire, daffodils can be adventurous plants in areas where they're happy. Interestingly, the Gowbarrow plants are exceptionally short, and reminiscent of a type of the species that grows in Spain. Maybe they had crossed a larger lake, thousands of years before.

It was not until two years later that William wrote his poem about their experience. And literal-minded sceptics (like myself, in those days) who learned of its initial inspiration in the excited detailing of his sister's diary, felt that not only did 'I wandered lonely ...' read by comparison like something in the poetry corner of a parish newspaper but it wasn't even true. William had been far from solitary on daffodil day. But that was the whole point. Lucy Newlyn, in her brilliantly empathetic study of the Wordsworths' working partnership (subtitled *All in Each Other*), imagines an occasion when the poem may have sprung to life. William, Dorothy and Mary are together, perhaps talking after supper, and something prompts Dorothy to read out her account of that April journey. It's a record of an occasion remembered fondly by both siblings – and gives pleasure to Mary too, who was not with them but, as William makes clear in his letters, contributed the 'the two best lines' – 'They flash upon that inward eye/ Which is the bliss of solitude'. The poem is a family creation. The 'I' is not literally William, wandering in solitude, hand pressed to forehead, but what Newlyn calls 'lyric utterance itself'. I can make sense of this by seeing it as the beginning of a classic Romantic trajectory, the abstract, aerial view of 'nature' suddenly grounded, brought back to real earth by the cheerful dance of the community of flowers, which reflected William and Dorothy's own comfortable friendliness that day.

I think the Golden Triangle's dillies brought me down to earth, too, but not because I was cheered by their animated dancing. In fact I don't remember them moving at all. Perhaps it was a windless day but I don't think that was the reason. I think I was relishing their *stationariness*, their vegetative love of their own *patria*. But they were beautiful earthly things too. Fran and I worked our way through the flowers' distinctive features: the different yellows of trumpet and ruff; the way the six petals of the latter were slightly cupped round the former; the pertness of them, the lean of one stalk away from another, as if they were already a posy; the slight silvering of the underside of the leaves. But in the end it was not so much their good looks that moved me as their presentness. I'm reluctant to say 'presence', with its hint of some spiritual aura, and mean just their 'being there', at that place and in that moment; present in the sense of 'all present and correct'. I was touched by the way they belonged to this spot, and that its human inhabitants had been contributors to that persistence – not as hubristic cultivators, managing and force breeding them, but as neighbours and companions, affectionately nudging the daffs along here and there (and not averse to making a few bob out of them) in a partnership of mutual benefit.

18

On Being Pollinated: Keats's Forget-Me-Not

LINNAEUS HAD HOPED TO MAKE his system of classifying plants according to their sexual organs more accessible by calling the stamens, the male parts, 'husbands', and the female stigmas 'wives' or 'brides'. So the *Enneandria* were described as 'Nine men in the same bride's chamber, with one woman' and *Adonis* as a mass orgy, with a hundred of each sex. Far from helping to convey his message, the vision of a kingdom of sexualised and, worse, licentious vegetables was too much for eighteenth-century sensibilities. The *Encyclopaedia Britannica* castigated Linnaeus: 'A man would not naturally expect to meet with disgusting strokes of obscenity in a system of botany.' The Revd Richard Polwhele agreed. His 'The Unsex'd Females' was a verse parody of Darwin, in which he lashed out equally at libertine vegetables and liberated women. He had particular venom for Mary Wollstonecraft, self-educated author of *A Vindication of the Rights of Woman* (1792) and a woman widely vilified for her affairs, figuring her as *Collinsonia* – 'two husbands in bed with one wife' – and enjoying a life of 'botanic bliss'.

It's uncertain whether any of the Romantics knew the full extent of plants' exchanges with insects, and what we now understand to be sexual congress by proxy. They were obviously aware that the two orders kept close company, and found the flower's gift of nectar to the bee richly metaphorical, analogous to their offerings of sweet scent and remembrance to humans. The pinpointed beauty of the forget-me-not (in Linnaeus's system the flowers are typified as having five men in bed

with four wives), the compelling fixity of its 'eye', the little yellow ring
at the centre of the corolla, made it a particular favourite. Coleridge
composed his poem 'The Keepsake' around it in 1802, recalling his love-
at-first-sight meeting with the lively redhead Sara Hutchinson (the elder
sister of Wordsworth's wife Mary) three years before:

> Nor can I find, amid my lonely walk
> By rivulet, or spring, or wet roadside,
> That blue and bright-eyed floweret of the brook,
> Hope's gentle gem, the sweet Forget-me-not!

In the second verse he fantasises a marriage with Sara (respect-
ably disguised under the name of Emmeline), and her stitching on a
silk sampler 'Between the Moss-rose and Forget-me-not – / Her own
dear name, with her own auburn hair!' The name forget-me-not hadn't
become current in Britain yet, and Coleridge felt obliged to add a slightly
pedantic gloss: 'One of the names (and meriting to be the only one) of
the *Myosotis Scorpioides palustris* [water forget-me-not] a flower from
six to twelve inches high, with blue blossom and bright yellow eye. It has
the same name over the whole of Germany (*Vergissmein nicht*) and, I
believe, in Denmark and Sweden.'

Keats wrote about the forget-me-not in 1818, also mentioning its eye.
He'd been swapping verses with his close friend and fellow poet John
Reynolds about, amongst other things, the allure of eye colour and the
potency of the colour blue. Blue – primary, primordial, reflected by sky
and sea long before the greens of life emerged – was seen as the most
transcendental of colours by the Romantics. John Ruskin pronounced,
'The blue colour is everlastingly appointed by the Deity to be a source of
delight.' (Modern social surveys suggest he was right, and that blue is a
strong candidate for the 'most favourite' colour.) Reynolds had written
a sonnet to Keats called 'Sweet Poets of the Antique Line' which ends
'Dark eyes are dearer far/ Than those that mock the hyacinthine bell'.
Keats responded with a sonnet subtitled 'The Answer ...', a hymn to the
colour of the hyacinthine bell. After rhapsodising on the blues of sky
and water, he ends:

Blue! – gentle cousin of the forest green,
Married to green in all the sweetest flowers –
Forget-me-not, – the blue-bell, – and, that queen
Of secrecy, the violet: what strange powers
Hast thou, as a mere shadow! But how great,
When in an eye thou art alive with fate!

Keats obviously means blue in human eyes in the last two lines. The
curious thing is that they would make equal sense if he were referring
to the power of blue in the flowers' 'eyes'. Thirty years previously two
German botanists, Joseph Kölreuter and Christian Sprengel, had worked
with forget-me-nots to unravel the seductive cues by which flowers
attracted insects to fulfil their vegetal destinies. Had Keats heard the
buzz?

Some sort of productive association between insects and flowers had
been intuited as far back as the Greeks. The Romans certainly understood the sexual nature of the pollination of date palms. Theophrastus
(circa 300 BC) talked of the need to bring the male flower to the female,
to ensure successful fruiting. But until the nineteenth century self-
pollination was commonly assumed to be the universal means by which
plants propagated themselves, and the role of nectar in this process was
often comically misinterpreted. One eighteenth-century theory saw it
as a kind of food for the plant, which might 'conveniently serve the same
purpose as white of egg', nourishing the seeds, and helping to 'make
them keep and preserve their vegetable quality longer'. The ever-imaginative Erasmus Darwin thought that flowers fed on their own nectar in
order to mature, so that they could 'become sensible to the passion, and
gain the apparatus for the reproduction of their own species'. From here
it was a short step to the conceit that the first insects had developed
from anthers which had 'by some means loosened themselves from their
parent plant, like the male flowers of *Vallisneria* [tape grass]; and that
many other insects in process of time had been formed from these'.

Philip Miller, a Londoner like Keats, and Director of the Chelsea

Physic Garden, made the first clear notes on insect pollination, which were written up by his friend Patrick Blair in 1721:

> [He] experimented with twelve Tulips, which he set by themselves about six or seven Yards from any other, and as soon as they glew, he took out the *Stamina* so very carefully, that he scattered none of the Dust, and about two days afterwards, he saw bees working on Tulips, in a bed where he did not take out the *Stamina*, and when they came out, they were loaded with Dust on their Bodies and Legs. He saw them fly into the Tulips, where he had taken out the *Stamina*, and when they came out, he went and found they had left behind them sufficient to impregnate these Flowers, for they bore good ripe Seed: which persuades him that the *Farina* may be carried from place to place by Insects ...

Miller's insights made little impact at first, but in 1750 Arthur Dobbs (of tipitiwitchet fame) made close observations of the flowers visited by bees in a hayfield, and examined the pollen loads carried back to the hive:

> Now if the Bee is appointed by Providence to go only, at each Loading, to Flowers of the same Species, as the abundant *Farina* often covers the whole Bee, as well as what it loads upon its Legs, it carries the *Farina* from Flower to Flower, and by its walking upon the *Pistillum* and Agitation of its Wings, it contributes greatly to the *Farina*'s entering into the *Pistillum* and at the same time prevents the heterogeneous Mixture of the *Farina* of different Flowers with it; which, if it stray'd from flower to flower at random, it would carry to Flowers of a different Species.

This concentration on similar blooms, commonly known as 'species constancy', is a matter of convenience rather than instinct for many non-specialised feeders. Even bees can be promiscuous feeders when opportunity arises, as a few minutes spent watching individuals in a mixed flower bed will demonstrate. But Dobbs had provided confirmation of Miller's broad ideas about the role of insects in pollination. The full and complex mechanisms of the nectar-for-sexual-services transactions were finally unravelled by Kölreuter and Sprengel in the last decades of the eighteenth century. Kölreuter established that nectar both attracts bees to flowers and is their high-energy fuel. (Pollen is used chiefly as a long-term food for the developing larvae.) Sprengel worked out the logistics of the harvest. In 1787, wondering what might be the purpose of the hairs on the base of the petals of blue wood cranesbill, he reckoned that since nectar was necessary to attract, and feed, pollinators, the hairs were a gauzy shower cap, serving to protect it from being spoilt by rain. The next year he studied forget-me-nots in depth, and recognised that the yellow ring (the iris of the flower's 'eye') acted as a honey guide, leading the visiting insects to the short tube at the centre of the sky-blue flower, and so to the nectar at its base. And, incidentally, to the pollen en route. Over the following years he distinguished four crucial components of the forget-me-not's floral structure: the nectary itself, which secretes the sugary juice; the nectar reservoir; the nectar cover, which protects it from rain; and the various devices which enable insects to locate the nectar – especially 'bee-line' markings and scent.

Forget-me-nots are pollinated chiefly by flies, including bee flies, whose appealing hovery flights in March are one of the most reliable signs of a warm spell. Experiments involving scent masking and fake flowers of different colours have shown that bee flies respond chiefly to scent, and secondly to blue coloration. Butterflies also occasionally visit forget-me-nots, and different species are highly individual in the senses they use to locate the nectar: small tortoiseshells don't respond to scent, but do to both the flower's yellow eye and its blue petals; peacocks are attracted by the scent and by blue and yellow; brimstones and other white family butterflies respond to the blue petals alone. Blue, and its shadings off into purple and ultraviolet, appears to be the 'favourite' colour in the insect world, too. Yellow is second, and is the partner of blue in greens. No insect has yet been found to have a receptor for red in its eyes, and colours at this end of the spectrum appear black or dark grey to them.

By the mid eighteenth century an understanding of co-evolution was emerging, a realisation that small changes in a flower's architecture or signalling abilities encouraged comparable adaptation in the insects that used it. Both parties benefited. The modified insect could continue to gather nectar, and the flower to have its pollen carried to

A bee fly nectaring on a forget-me-not.

other individuals, ensuring a constantly shifting and resilient genetic base. The transport of pollen probably developed as a spin-off from insects eating primitive self-pollinating flowers. But floral diversity, as we understand it, evolved chiefly as a result of the mutually beneficial relations of these two classes of organism.

Did Keats suspect any of this back in the early years of the eighteenth century, and if so how would it have sat with his apparent hostility to reductionist science? He'd been a florophile since he was a boy in the villages that made up eastern London. He'd foraged in open fields around Edmonton, watching birds, climbing elms and gathering nettle leaves to hide in his brothers' beds. His medical notebooks from ten years later are decorated with small sketches of flowers, including the pages devoted to Astley Cooper's lecture on the structure of the human nose. His poems are as full of floral decoration as the interiors of Gothic churches. His last letter, written just before he died on 23 February 1821, aged just twenty-five, has a final flourish: 'O! I can feel the cold earth upon me – the daisies growing over me – O for this quiet.'

Three years earlier he found in the reality of the flower a perfect model for his idea of 'Negative Capability' – a kind of creative suspension of certainty, an attentive indolence, the 'wise passiveness' advocated by Wordsworth. One beautiful spring morning in late February he sat in his rooms in Hampstead and wrote a cheering letter to John Reynolds, then passive from chronic illness, not philosophical choice.

> It has been an old Comparison for our urging on – the Bee Hive – however it seems to me that we should rather be the flower than the Bee – for it is a false notion that more is gained by receiving than giving – no, the receiver and the giver are equal in their benefits. The flower I doubt not, receives a fair guerdon from the Bee – its leaves blush deeper in the next spring – and who shall say between Man and Woman who is the most delighted? Now it is more noble to sit like Jove than to fly like Mercury – let us not therefore go hurrying about and collecting honey bee-like, buzzing here and there impatiently from a knowledge of what

is to be arrived at; but let us open our leaves like a flower and be passive and receptive – budding patiently under the eye of Apollo and taking hints from every noble insect that favours us with a visit – sap will be given us for Meat and dew for Drink.

I was rather taken, on first reading, with Keats's chirpy but subtly subversive praise of the flower – or at least of laid-back flower power as a model for living: he was giving his sick friend a pick-me-up, a get-well note, not an ecology lesson. But I find it hard to credit that his allegory of human creativity and reciprocal relationships isn't based on the mutuality of pollination, and suspect Keats was aware that more was going on between flower and bee than an exchange of nectar and a debate about utilitarian values. There are innuendos of sex and intimations of ecology in his talk of 'budding', and flowers 'taking hints' from insects, and in the notion of the flower receiving 'a fair guerdon' (a reward) in return for its nectar, something which effects its future growth. Keats had read broadly in science for his medical studies. His suggestion that males and females might have equal 'delight' in congress was as scientifically precocious as it was socially progressive, and hints that he may have been aware of contemporary notions of insect pollination, which allowed flowers to have, as it were, sexual intercourse by proxy.

But Keats's views on science remained ambivalent. Would the knowledge that flowers had a *function*, beyond providing metaphors for lotus eating and creative passivity, put him on the side of Wordsworth's dictum in 'The Tables Turned': 'Sweet is the lore which Nature brings;/ Our meddling intellect/ Misshapes the beauteous forms of things:–/ We murder to dissect'? Keats had, with reluctance, done his own dissection at medical school, and had notes on the surgeon John Hunter's physical unweaving of the electric eel in his attempt to discover the source of its vital spark. Yet in some very well-known lines from his long poem 'Lamia' (1820) he repeats his attack on Newton's dissection of the rainbow:

Do not all charms fly
At the mere touch of cold philosophy?
There was an awful rainbow once in heaven:

We know her woof, her texture; she is given
In the dull catalogue of common things.
Philosophy will clip an Angel's wings,
Conquer all mysteries by rule and line,
Empty the haunted air, and gnomed mine –
Unweave a rainbow ...

'Lamia' is a complicated narrative poem, in which a terrifying serpent, or rather an extraordinary snake-like chimera – 'Vermilion-spotted, golden, green, and blue;/ Striped like a zebra, freckled like a pard' – is transformed into a woman, and then falls to her true serpent form again when challenged by the fierce gaze of the rational philosopher Apollonius. Keats's reference to reason's power to 'Conquer all mysteries by rule and line,/ Empty the haunted air, and gnomed mine' is deliberately equivocal. 'Cold philosophy' could be protective as well as disenchanting, liberating us from the dark powers of unreason.

Within fifty years science would begin to undermine Keats's pleasant conceit of the gently passive and receptive nature of flowers. Since Charles Darwin's unravelling of the pollination procedures of primroses and orchids, flowers have been shown to be as potentially busy and impatient as bees. Indeed, they could be said to make the first move in most flower–insect relationships – firing smell pulses, electrostatic charges, reflecting sound waves to tempt pollinators towards them, then imprisoning them with fantastic arrangements of trapdoors, one-way tunnels and chemical handcuffs. Insects, for their part, have been shown to have developed discrimination and a kind of instinctive patience in their choice of nectar partners. I think Keats might have enjoyed the deeper layers of reciprocity now known to exist between flower and bee, and that *choices* (however circumscribed) are being made by the insect. Modern writers such as David Rothenberg and Michael Pollan have suggested that there are analogies with our aesthetic sense in this process: we too are drawn to flowers whose appearance gives us some kind of affective reward. Pollan isn't quite right in saying that the flower acts as a metaphor to the insect; but it is a signifier, a dazzling multi-sensorial advertisement for what lies hidden beneath.

For us, flowers are both metaphors and signifiers. The eye of *Myosotis* says 'forget-me-not' and bee fly, lover and poet obey. And the Romantic imagination, inspirited by nature, pays it back in a fair guerdon of art.

The last of the nineteenth-century Romantics, John Ruskin, was appalled by any suggestion that the 'beauty' of flowers existed chiefly to delight insects and assist plant reproduction. He had written with enormous understanding about the elegant forms and delicate engineering of leaves but was repelled by the idea that they were built just so for the business of photosynthesis. This is a beautiful and magical process to most modern minds, but to Ruskin it was offensive, reducing leaves to the status of 'gasometers'. Similarly, his sensuous and evocative descriptions of flowers, which have no equal in nineteenth-century prose, ran in parallel with a belief that they existed purely for the edification of the human soul: 'The perception of beauty,' he wrote in *Proserpina: the Studies of Wayside Flowers*, 'and the power of defining physical character, are based on moral instinct.' Ruskin didn't deny that the forms of plants could be functional, but denied that 'beauty' could be an objective measure of the grace and elegance with which they existed on their own terms, and inside their own living communities. He believed it could only ever be a value judgement by humans with the divinely endowed gift of seeing moral purpose and design in nature. Which is why he took the unconventional view that the flower, not the seed, was the ultimate achievement, the final purpose of a plant's life:

> [T]hese are the real significances of the flower itself … It is the utmost purification of the plant and the utmost disciple. Where its tissue is blanched fairest, dyed purest, set in strictest rank, appointed to most chosen office, there – and created by the fact of this purity and function – is the flower. But created, observe, by the purity and order, more than by the function. The flower exists for its own sake, not for the fruit's sake. The production of the fruit is an added honour to it – it is a granted consolation to us for its death. But the flower is the end of the seed, – not

the seed of the flower ... It is because of its beauty that its continuance is worth Heaven's while.

Ruskin died in 1900. Had he lived for another quarter of a century he would have been able to confront a considerable challenge to his aesthetic theory: an orchid which flowered entirely underground and out of sight. In 1928 a Western Australian called Jack Trott noticed an odd crack that had appeared in his flower bed, and a sweet scent percolating from it. He scraped away a few inches of soil and found a small, pale pink efflorescence buried underneath. It proved to be an entirely new species, later named *Rhizanthella gardneri*, the only member of the orchid tribe to live out its entire existence underground. Not able to receive any energy from the sun, it has a symbiotic relationship (via a mycorrhizal fungus) with the evergreen broom honeymyrtle, or broombush, *Melaleuca uncinata*. Its blooms are small, about an inch across, but perfectly and strangely formed. The pale sepals enclose up to ninety tiny, dark maroon orchid flowers, like caviar in a cup. They are pollinated by underground insects such as termites, presumably attracted by their sweet smell. The whole flowering apparatus's visual beauty is irrelevant and unwitnessed (except, presumably, by God), and only Ruskin himself could pronounce on whether this disqualifies it from the purity of flowerhood altogether or makes it a sublime example of floral beauty existing for its own sake.

NEW LANDS, NEW VISIONS

IF THE EIGHTEENTH AND NINETEENTH CENTURIES were a time when the images and meanings of familiar plants were reframed, the prodigious new growths being discovered in European colonies in the tropics and even in the Mediterranean demanded even greater intellectual and emotional readjustment. These strange and luxuriant growths strained imaginations, as the desire to subsume them as imperial wealth jostled with the curiosity to understand them as living things. The Royal Botanic Gardens at Kew were the focal point of this international vegetal conversation. Kew sponsored expeditions, acted as a clearing house for economically promising species, and archived both living plants and a huge range of their artistic images. It became a kind of man-made ecosystem, a microcosm of all the relations possible between plants and people.

I've been fortunate to work with Kew several times over the past decades. I reported my friend David Nash's 2013 sculpture exhibition, which filled the site with compelling works in wood commenting on everything from the hubris of colonialism to the geometry of plant form. In 1987 I recorded how the gardens coped with a massive piece of natural wood sculpturing, the Great Storm of 17 October. The appeal of the visual is ubiquitous and ambiguous at Kew, as it is throughout the universe of plants. Earlier that year I'd spent some weeks immersed in its archive of more than a million illustrations. It's an extraordinary collection, containing elaborate 'florilegium' pieces by professional artists, pencilled marginalia by botanists 'in the field', watercolours by colonial governors' wives with time on their hands, and the painstaking works of the artists sent out with Kew's plant-collecting expeditions over three centuries. These, especially, make a fascinating counterpoint to the records of the expeditions' scientists. They often had social backgrounds different from those of the expedition leaders. Their encounters with plants were also mediated by the business of painting itself, which demanded an intimacy which was not always part of the work of scientific collecting. The cultural historian Mary Louise Pratt has written about how, in the process of imperial expansion, 'one by one the plants' life forms were to be drawn out of the tangled threads of their life surroundings and rewoven into European-based patterns of global unity and order'. She continues, in her book *Imperial Eyes*: 'The (lettered, male, European) eye that held the system could familiarise ('naturalise') new sites/sights immediately upon contact, by incorporating them into the language of the system.' Yet botanical artists were by no means all lettered, male and European. Indigenous artists, working-class illustrators, self-financed women all painted plants in the colonies. They often had their own agendas, and far from 'immediately' incorporating 'new sites/sights' into 'the system' had to broaden their visual language to make sense of hard-to-comprehend new floras. This was a response which was echoed a century later in the naturalistic plant painting of the Impressionists, which I touch on at this end of this section.

The natural-history artist who broke the mould in the late seventeenth century was Maria Sibylla Merian, born in Germany to a respectable and religious family of Swiss publishers and artists in 1647. The focus of her fascination was always insects, but the vegetable food they depend on is rarely out of the frame. She was intrigued by their metamorphoses, from egg to leaf-eating caterpillar to nectar-drinking butterfly. Despite her orthodox upbringing her life went through dramatic metamorphoses of its own. In 1685 she fell out with her family, left her husband, and converted to the Labadists, a radical Christian sect. Five years on she dropped out again, and Kim Todd's empathetic biography *Chrysalis* suggests that she intuited some resonance between her own hermetic existence and the disconnected lives of insects as portrayed in most contemporary illustrations. They were wrenched out of their living contexts, robbed of development and ecological relationships.

In the 1690s Merian made a decision to go with her daughter to the Dutch colony Surinam, to paint first hand the lives and transformations of the creatures whose inert remains were increasingly appearing in the collections of returning explorers. She found the native Amerindians and the African slaves more helpful than the European planters. They took her exploring in the rainforest, explained the traditional uses of plants to her, and brought her maggots and chrysalises. She sketched from life, and later worked up her drawings into paintings on vellum. Even then, she seemed close to the marrow of Amazonian life. Her insects and plants are telling stories of interdependence, of the value that is contained within their own circle of existence. But after two years she was unable to bear the tropical heat any longer and returned to Amsterdam, where her *Metamorphosis of the Insects of Suriname* was published in 1705. Its most extraordinary illustration, challengingly gruesome and bordering on the surreal, is 'Spiders and ants on a guava tree'. This is an ecological portrait out of Hieronymus Bosch. A giant bird-eating tarantula is mantling a dead hummingbird. It has one leg in a nest full of eggs (a detail which, to judge from its odd stiffness and sense of scale, Maria did not paint in the wild). The guava has ripe fruit, but moth-eaten leaves, the remains of which are being ferried about the twigs by ants. Another species of spider is catching ants in her web, while more ants are attacking a striped insect in the foreground. Life goes on, and goes round. It is an exaggerated picture, and some way out of proportion, but gives a sense of the alien and dynamic life of the tropics.

Maria Merian was exceptional in that she was a freelance, working for herself. But her experience as an artist in exotic regions was shared by the great numbers of botanical painters and illustrators who were employed by governments and botanic gardens – for example, Sydney Parkinson, working for Joseph Banks in Australia, Ferdinand Bauer

Maria Sibylla Merian's phantasmagoric painting of the ecology of a guava tree, made in Surinam, *c.* 1701–5.

in the Mediterranean, Francis Masson in South Africa. The way they painted was challenged and changed by the radical forms and ways of life of the new plants they encountered.

19

Jewels of the Desert: Francis Masson's Starfish and Birds of Paradise

FOR BRITAIN'S FIRST OFFICIAL roving plant collector, Francis Masson, it was the bizarre plants of the southern African deserts that transformed his vision. Masson was born in Aberdeen in 1741 and brought up to work as a gardener, and like many other Scots of his time decided to seek his fortune down south. He got a job as an under gardener at Kew, where his enthusiasm for botany soon caught the eye of his employers, and in particular the director, Joseph Banks, the Svengali of eighteenth-century botany. Ever since his tantalising first visit to the Cape of Good Hope, Banks had wanted to send an envoy to southern Africa, to gather seeds and specimens for Kew. An artisan who knew his place fitted the bill perfectly. Banks was later to describe Masson as 'a young gardener who has no education beyond his line of life', so that there would be no challenging the agenda or morals of the management. Perhaps neither man knew at that point that Masson had a natural talent for flower painting, and that – male, European but no lackey of the system – he had a latent sympathy for the cultures and ecosystems he was about to visit.

Masson set out on his exploration from Cape Town in December 1772, and over the next two years kept a methodical journal. He records the crops grown by the Dutch farmers, the local hunting customs and the wild animals of the plains. He saw lions, elephants and zebras. He laments the demise of the hippopotamus, once abundant in all the large rivers but almost exterminated since the arrival of Dutch settlers. Like most early colonial explorers, he first needed to see the landscape through a

European lens to make sense of it in his imagination. But reflecting on the fortunes of a country whose invaders had largely treated its natural wealth with imperious negligence, his conclusions didn't follow the usual colonial line:

> This tract of country has afforded more riches for the naturalist than perhaps any other part of the globe. When the Europeans first settled there, the whole might have been compared to a great park, furnished with a wonderful variety of animals ... but since the country has been inhabited by Europeans, most of these have been destroyed or driven away.

At least the flora seemed unspent and unexpected. Masson found species which were to become favourites in European gardens: ixias, gladioli and irises in the valley grasslands, proteas on the skirts of the mountains, heathers on the crags. Yet it was the dry and seemingly barren sands of the western coast, known as the Karoo, which captivated him. This is in southern Africa's equivalent of the 'Mediterranean biome', and he found the climate hot and 'dismal'. He was fascinated by the range of succulent plants – 'endowed by nature, as the camel is, with the power of retaining water' – that were able to flourish in this inhospitable wasteland. He records his arrival in the Karoo on 20 November 1773:

> At night we got clear of the mountains, but entered a rugged country, which the new inhabitants name Canaan's Land; though it might be called the Land of Sorrow; for no land could exhibit a more wasteful prospect; the plains consisting of nothing but rotten rock, intermixed with a little red loam in the interstices, which supported a variety of scrubby bushes, in their nature evergreen but, by the scorching heat of the sun, stripped almost of all their leaves. Yet notwithstanding the disagreeable aspect of this tract, we enriched our collection by a variety of succulent plants, which we had never seen before, and which appeared to us like a new creation.

Amongst these were the strange and almost unknown stapelias which were to form the subject of his best paintings. Stapelias look like cacti, but are related to subtropical vines (and to the American milkweed, host plant of the monarch butterfly). They have fleshy, toothed, photosynthesising stems, whose high water content is an adaptation to living in desert conditions. Their five-petalled flowers, sharp cut and symmetrical, have been called 'starfish flowers'. They're covered in fine hairs and can be as ornamented as embroidered fabrics. *S. ocellata*'s have the fine speckle of a wildcat's skin; *S. reticulata*'s are like Fabergé eggs, or sea urchins – red domes inside calyces patched with orange; *S. revoluta*'s are fully reflexed, and have the look of tiny velveteen jellyfish pulsing from the stem. (Masson doesn't mention that the starfish blooms stink incongruously of dead meat, to attract pollinating carrion flies.)

What is remarkable is how this inexperienced artist, on his first trip overseas, and nurtured by the contrived and prettified plant portraits of Pierre Redouté and Georg Ehret, is able to capture the weirdness of this new flora, and to convey visually something essential about how it functioned in a profoundly hostile environment. The succulence of the stapelias, which helps them hold stores of water in drought conditions, is one of the strategies that marine plants – like the samphire family – employ to survive the freshwater-sapping effect of the sea. Masson intuits this, sensing the convergence of plant form in the two habitats. A browse through his stapelia paintings is a submarine experience, a journey through a terrestrial reef. He paints the stems heavily, using shadow to give an almost impasto effect which emphasises their fleshiness. Some are as stiff and glaucous as coral, some possess the languidness of oarweed. The flowers have the look of exotic sea creatures, clinging precariously to the wavy stems, and he paints their hairs – also a water-conserving adaptation – individually. All that is missing from his pictures is a sense of the plants in their habitat – a type of botanical painting that wouldn't really begin for another half-century. If he had included this, we would see one other strategy for moisture conservation. Despite being supremely adapted physically to surviving in fierce dry heat, stapelias opt to grow, whenever it is possible, nestled in the cooling shade of desert shrubs.

Masson's upbringing may have helped him empathise with these organisms which so radically contradicted the European conviction that

Stapelia Gordoni.

Published as the Act directs June 30.1795. by F.Masson.

Stapelia gordoni, by Francis Masson, from *Stapelia Novae*, 1796.

rich plant growth depended on fertile soil. He would have been familiar with one kind of wasteland vegetation from the hills round his Aberdeenshire home. But by the side of the spangled tuftings of the Karoo, his home moors must have seemed like a monoculture. In his youth he would have been aware of the depauperisation of Scotland's upland flora as a result of the Highland Clearances. Even as far south as Aberdeenshire subsistence farming was being replaced by sheep herding on a vast scale. Masson recognised that an almost identical process was beginning to affect the Karoo.

The local tribespeople, the Khoikhoi (then called Hottentots by Europeans), were nomadic herders, grazing their Nguni cattle on common ranges and practising a variation of transhumance in which they and their stock migrated to more richly vegetated areas as and when rainfall or the seasons dictated. The growth of the stapelias and other desert natives (most of which are perennial and can withstand grazing) is highly dependent on rainfall, so this system made for a sustainable balance between plants and animals, with the nomadic graziers taking seasonal and sometimes opportunistic advantage of local conditions.

When the Dutch settlers arrived they adopted the Khoikhoi's herding strategies for a while. But when Masson was there in the late eighteenth century, they were beginning to create fixed, private grazing areas of the kind that characterised north European stock-raising farms, and which were quite unsuited to arid habitats. Settlement around private water sources meant that the grazing orbits shrank dramatically and became more intensively used. Livestock was moved from rangeland to water source to kraal on a daily basis, partly to provide some protection from predators. The concentrated movement of animals created erosion, and the fixed kraals soon became barren areas of windblown sand.

Masson travelled through similar terrain the following September, and a conversation with a 'peasant' (presumably a Dutch farmer) hints at the intensity of the grazing:

> 26 October 1774. The sterile appearance of this country exceeds all imagination: wherever one casts his eye he sees nothing but naked hills without a blade of grass, only small succulent plants ... The peasant told

us, that in winter the hills were painted with all kinds of colours; and said, it grieved him often, that no person of knowledge in botany had ever had an opportunity of seeing his country in the flowery season. We expressed great surprise at seeing such large flocks of sheep as he was possessed of subsist in such a desart; on which he observed, that their sheep never ate grass, only succulent plants, and all sorts of shrubs, many of which were aromatic ...

[31 October] This desart is extensive and ... [t]here still remains a great treasure of new plants in this country, especially of the succulent kind, which cannot be preserved but by having good figures and descriptions of them made on the spot; which might be easily accomplished in the rainy season, when there is plenty of fresh water every where. But at this season of the year, we were obliged to make the greatest expeditions to save the lives of our cattle, only collecting what we found growing along the road side, which amounted to above a hundred plants, never before described.

It was typical of Masson's personality to put animal welfare above professional ambition.

Masson drew a few wild specimens on his field trips, but most of his work is based on plants he had transplanted to his garden in Cape Town, where he'd settled in 1786. 'The figures', he wrote, 'were drawn in their native climate, and ... they possibly exhibit the natural appearance of the plants they represent, better than figures made from subjects growing in exotic houses can do.' Masson was being typically modest. His collection of paintings, published as *Stapelia Novae* in 1796–7, catch the exhilaration and optimism he felt at this blooming of the desert.

The Karoo's own future was to be of the kind familiar in so-called wastelands. The fighting of the Boer War raged across it at the dawn of the twentieth century, and in the early years of the twenty-first it was announced as a major site for the establishment of fracking in southern Africa.

Masson himself, for all his sympathy with the indigenous people and animals and plants of the Karoo, was not entirely immune to imperial sentiment. On a trip back to Britain in 1773 he brought with him the first specimen of the bird of paradise flower, one of five species of *Strelitzia* which grow on river banks and in damp scrub in Cape Province. No flower as lavishly coloured or extravagantly shaped had been seen in England before. It is over six feet tall, and the flowers stand high above the leaves on the tips of the long stalks like exotic cranes. The hard, pointed sheaths, washed with pastel pinks and mauves, are held at right angles to the stem, in the manner of a bird's head and beak. From this sheath the flowers emerge laboriously, each one taking a week to unfurl. They comprise three flamboyant orange sepals and three diminutive blue petals, together giving the appearance of an ornamental crest. The flowers imperious, almost heraldic form, led to its dubbing as *Strelitzia reginae*, in honour of Charlotte of Mecklenburg-Strelitz, queen consort of the United Kingdom, and painters from Kew and the Court circle set themselves the task of loyally capturing its image on paper. The Austrian botanical artist Franz Bauer devoted an entire volume to it (*Strelitzia depicta*, 1818). He caught well enough the sunset blues and oranges of the flowers, but their plastic sheerness, echoed more strongly in the surface texture of the sheath, proved beyond his skills. As it did for Masson, for all his later ability to evoke novel plant forms.

One other notable feature of the floral bird of paradise, undrawn and unremarked by Masson, is that it is a magnet for real birds. Sunbirds come to sip nectar from the plant, perching on its flowers to gain access to the honeyed inner chambers. Their weight causes the petals to open and cover their feet with pollen, so enlisting the birds as pollinators in exchange for their nectarous harvest. As we've seen, the mechanisms of plant pollination by insects were just being unravelled during these first decades of the nineteenth century. The idea that exquisite birds might also be functionaries, as well as diners on honey, was too radical by far.

20

Growing Together: The East India Company's Fusion Art

INDIA HAD BEEN COLONISED earlier than southern Africa, but at the beginning of the nineteenth century its resources of timber and potential food and medicinal plants had barely been surveyed, let alone brought into the service of Europe's empires. It had one other boon not common elsewhere in the tropics: an educated workforce with a written culture. One East India Company manager became quite breathless when he contemplated the potential of this untapped resource: 'What a vast field lies open to the botanist in that boundless country. How many unemployed individuals are there whose leisure hours might be agreeably, usefully and profitably employed in this pursuit! ... Great God, how wonderful, how manifold are Thy works!' This vision of the subcontinent's unemployed finding job satisfaction and possibly salvation in Flora's arms wasn't entirely typical of the East India Company's mission. It had arrived in India in the early seventeenth century, and acquired a degree of power remarkable for what was notionally a purely commercial concern. It had a virtual monopoly on the exploitation of the country's economic resources, and in many regions acted as a de facto government. Where it was in its own interest, it suppressed or appropriated key local industries.

But local botanical knowledge was another matter. The Company calculated that the Indian flora almost certainly contained plants of unrealised economic importance, and that local plant lore – and indigenous scientific knowledge – might be a short cut to discovering these. It began methodical survey work in the middle of the eighteenth century, and in 1787 set up a botanic garden in Calcutta. The first superintendent, Lt Col. Robert Kyd, was hard-headed but diplomatic about the Company's role. He saw the garden's function as 'not for the purpose of collecting rare plants as things of curiosity or furnishing articles for the gratification of luxury, but for establishing a stock for disseminating such articles as may prove beneficial to the inhabitants as well as the natives of Great Britain, and which ultimately may tend to the extension of the national commerce and riches'.

Meanwhile, in the unexplored south-east coastal region known as Coromandel, a young Scots physician and botanist had already begun an impressive documentation of the region's useful plants. William Roxburgh had joined the Company as a surgeon in 1776, but spent much of his time studying local plants. By 1789 he had abandoned medicine, and taken up the post of Company botanist for the region around Madras. With the job he also acquired a project started by his predecessor, Patrick Russell. Russell had dreamed of compiling an economic flora for Coromandel, and his outline had received the all-important approval of Joseph Banks. Roxburgh picked up the threads of the proposed book on Russell's retirement, with a good proportion of the material already prepared. He had his own field notes and a portfolio of illustrations by an Indian artist, whom he'd kept 'constantly employed in drawing plants, which he accurately described, and added such remarks on their uses as he had learned from experience or collected from the natives'.

Roxburgh eventually accumulated more than 2,500 paintings for the Company, of which Joseph Banks picked 300 for publication. They were published in twelve parts between 1795 and 1820 as *Plants of the Coast of Coromandel*, which remains one of the most extraordinary collections of flower paintings to have been published in Britain. Banks thought it was the finest Indian flora to have appeared in Europe. What makes the illustrations exceptional is their quality as cultural hybrids. The Company would have preferred accurate and unornamented field-guide illustrations, a crib for collectors and prospective cultivators. What they got was an exotic fusion of European precision and Mughal stylisation that revelled in the pure patterning of plants.

Sappan, painted by an unknown Indian artist for the East India Company and published in William Roxburgh's *Plants of the Coast of Coromandel*, 1795.

There was a long tradition of flower painting in Mughal culture, chiefly of delicate miniatures built up by layer upon layer of brilliant body colour. The paintings were finished by the use of very fine brushes, which were drawn across the paint to add texture and surface detail. In this way it was possible to suggest the lustre of petals or the leatheriness of leaves. By modern standards traditional Mughal flower paintings are exquisitely detailed and highly successful at catching the jizz of a plant – not by impressionism so much as a kind of hyper-realism. But the East India Company regarded them, unmodified, as too obviously decorative. They lacked the literalism and austere clarity of line that had become customary in European plant illustration, and which were regarded as 'proper' for scientific representation. Nor did they employ techniques – the use of perspective, for example – which in the West were regarded as essential for highlighting a plant's characteristic features, and providing guides to reliable identification.

So, as part of the process that led eventually to the chimerical style of *Coromandel*, Company officials had begun training Indian artists in European techniques. One of the models offered was James Sowerby's work for the *Flora Londinensis*, with its fastidious attention to detail and to all the plants' internal and external structures. They introduced the local painters to the subtleties of watercolour, and suggested how their own bright pigments might be muted for more restrained British eyes. The identity of the Indian artists isn't known, except that they were mostly Hindus, and may have included artists such as Haludan, Vishnu Prasad and Gurudayal, who are known to have worked for the East India Company. They seemed happy to go along with this new approach: painting for the Company was at least regular work, even if they were paid a pittance. But though they succeeded in achieving the kind of accuracy demanded by the Company, the habits and traditions of their culture couldn't easily be suppressed, and the house style that evolved – which came to be called 'Company Art' – is uniquely cross-cultural. On the surface the paintings are neat, comprehensible, even diagrammatic where necessary. But working inside these conventions, and without ever falsifying the plant's detailed identity, the painters made compositions, relishing the contrast of colours and shapes, looking for suggestions of order, portraying an intriguing surface detail so sharply it looks as if it has been etched. They were, unconsciously, following a long and global tradition here, stretching from the late Palaeolithic, through the lush ornamentation of Gothic carving in Europe and Chinese landscape painting, where the inventiveness and caprice of plant growth are celebrated.

The water chestnuts, *Trapa* spp., are a family of floating aquatic plants whose tuberous roots, rich in starch and fat, are a staple in Asian cooking and familiar in oriental dishes in the West. The eighteenth-century Hindu painter of *Trapa* ignores the all-important root, but is carried away by the floatiness of the leaves. Shaped like spades and edged with dark hatching, they are fanned out from the stem like a hand of cards. The lagerstroemias, or crape myrtles, are now popular garden shrubs, but they're seen afresh in the *Coromandel* painting, where the artist has emphasised the cut-and-crimped paper appearance of the flowers, and elegantly arranged them as a chaplet round the shoots, though he's rather awkwardly twisted one of the leaves round so that the underside is shown. A reluctance to be as literal minded as their teachers hoped was common. Shading, intended to give a sense of depth, sometimes appears on the wrong side, and backgrounds can be so intense that they overpower the main subject. The illustration of sappan (*Caesalpina sappan*, source of a valuable red dye), which positions the small yellow flowers against the ferny leaves, has the look of a piece of jazzily printed fabric, but hardly shows off the flowers to best advantage. Large leaves especially were apt to be painted in flat, unmediated greens, with an appealing eggshell finish but not much leafy realism. 'Most abominable leaves for which Master painter shall be duly cut,' reads a severe Company jotting on the back of one of the original pictures. Yet Maria Graham, sister of the Professor of Botany at Edinburgh, watched some of Roxburgh's artists at work in 1810, and thought their paintings 'the most beautiful and correct delineations of flowers I ever saw'.

The lasting impression these paintings give, beyond the inventiveness of their patterning and layout, is their sense of light and heat. Sometimes the lack of perspective and shadow seems odd to eyes

Lagerstroemia Reginæ

(Roxburgh)

The crape myrtle *Lagerstroemia speciosa*, another example of 'Company art' from *Plants of the Coast of Coromandel*.

conditioned by northern climates and landscapes, but the pictures have the feel of plants painted in the sunshine. This was a revelation to painters used to European gloom. When William Hodges, landscape painter on Cook's second Pacific trip, travelled in India, it was the light which impressed him above all: 'The clear, blue, cloudless sky, the polished white buildings, the bright sandy beach, and the dark green sea, present a combination totally new to the eye of an Englishman, just arrived from London, who accustomed to the sight of rolling masses of clouds floating in damp atmosphere, cannot but contemplate the difference with delight.' Perhaps this was as important a legacy of colonial botanical art as the new species it brought into the northern consciousness – the suggestion of a different kind of biological energy. A similar process was to happen in Europe half a century later, when the dazzle of the Mediterranean flora helped give birth to modern art.

21

Chiaroscuro: The Impressionists' Olive Trees

IN EUROPE THE PRESENCE OF OLIVE TREES defines the Mediterranean region, with its unique climate and intense light. The line enclosing the area where the tree will prosper closely follows the contour joining places with a mean February temperature of 7 °C (45 °F). It passes through central Spain, southern France, the Italian lowlands, southern Greece and the islands, the Middle East and back through northern Tunisia and Morocco. It also marks the boundary of southern, 'Latinate' landscapes. To cross over the olive line, as many travellers and painters did in the eighteenth and nineteenth centuries, is to pass from the bright and sappy greens of the deciduous north to the silvers and grey-greens of the south. Olives, massed in groves, are the top layer of this colour shift, overlaying the rosemary and cistus and lavender of the garrigue, and the overall tonal change is a consequence of adaptations in plant anatomy. The Mediterranean climate is typified by hot, dry summers and mild, moist winters. Conservation of water is an imperative for all plants. Many species are evergreen and can continue photosynthesising through the cooler months. Their leathery and resinous leaf surfaces act like oilskins, reducing water evaporation. Grey and silvery tones are usually due to dense hairs, which have the same effect. Olives employ both devices, having leaves with blue-green upper surfaces and grey-haired undersides. In windy conditions the leaves close slightly, protecting the transpiring upper surfaces from drying out. Even in a breeze olive trees seem to shimmer in alternations of green and blue and silver, the undersides of the leaves showing matt grey, and the angled, resinous upper surfaces reflecting the sun in flashes of burnished bronze. The leaves themselves are rigidly fixed to the branches, but these are as flexible as willows so that as they move their shadows flicker through the interior of the tree in a show of natural chiaroscuro.

If the olive's fruit and oil are the Mediterranean's great gifts to world cuisine, its foliage has helped shape modern Europe's visual sense. In his 1934 essay 'The Olive Tree' Aldous Huxley calls it 'the painter's own tree' and suggests that it defines and shapes the look of the south:

> The olive tree is, so to speak, the complement of the oak; and the bright hard-edged landscape in which it figures are the necessary correctives of those gauzy and indeterminate lovelinesses of the English scene. Under a polished sky the olives state their case without the qualifications of mist, of shifting lights, of atmospheric perspective, which give to English landscapes their subtle and melancholy beauty.

The first detailed painting of the leaves in eighteenth-century botanical art was by the young Austrian artist Ferdinand Bauer (Franz's brother). He had travelled to the eastern Mediterranean at the end of the eighteenth century with John Sibthorp, Professor of Botany at Oxford. Their project was to create a sumptuous *Flora Graeca*, a Mediterranean corollary of the extravagant tropical floras of the colonial era. (Only twenty-five copies were finally published in 1840, at a colossal price, but the whole venture helped open up the plant life and landscapes of the region.) Bauer's olive leaves are painted in watercolour. There are just two, partially overlaying each other, the top one showing its pale silvery underside, the lower its dark green upper surface. They are exact and far from impressionistic, but they understand the trick olives play with light and which would help change the whole direction of painting nearly a century later.

Cézanne painted them first, posing brooding olives in the foreground of his obsessive pictures of Mont Sainte-Victoire in Provence. Van Gogh, desperate to 'feel the whole of the country' ('isn't that what distinguishes a Cézanne from anything else?'), travelled south in 1889, and wrote to his brother: 'Ah, my dear Theo, if you could see the olives at this moment ... The old silver foliage and the silver-green against the blue. And the orange-hued turned earth. They are totally different from what one thinks in the north ... The murmuring of an olive grove has something very intimate, immensely old. It is too beautiful for me to try to conceive of it or dare to paint it.' But dare he did, and in the late autumn painted four large canvases, one after the other, in which the olives are shown at different times of day, in different moods, against different vistas. 'He painted them,' his biographers write, 'with their emerald foliage flaming like cypresses, and with their silvery under-leaves sparkling like stars.' He created a total of eighteen olive canvases in his life. Renoir, too, was transfixed by the light on olive foliage. 'It sparkles like diamonds,' he wrote in his journal one afternoon. 'It is pink, it is blue, and the sky that plays across them is enough to drive you mad.' Huxley, writing fifty years later, believed he could see how the Provence olives helped give birth to Impressionism. The tree 'sits lightly on the earth and its foliage is never completely opaque. There is always air between the thin grey and silver leaves ... always the flash of light within its shadows' and Van Gogh's blues and Renoir's pinks too, despite the fact that 'no olive has ever shown a trace of any colour warmer than the faint ochre of withering leaves and summer dusts'.

I have a modern watercolour of an olive grove in Extremadura in central Spain, by my late friend David Measures. It's post-Impressionist in a literal sense, and rather abstract in its use of colour. The olive trunks are striped with turquoise and jade. The foliage picks up flecks of orange and pure white. I went to stay at the farm where it was painted one spring and saw that the colours weren't abstract at all. Their random fluttering picked up colours from elsewhere in the grove, the scarlet poppies growing beneath the trees, the rufous plumage of feeding hoopoes, scraps of burnished sky. Reflected off the silver leaves, they left an afterglow in the eyes.

Renoir bought his own olive grove in 1907. He was sixty-six years old and severely arthritic, and had taken to spending the winters in Provence. In 1904 he discovered Les Collettes, a rundown farm in Cagnes with an ancient olive grove attached to it. He learned from a

'The Olive Grove', Vincent van Gogh, 1889.

villager that the widow who owned the property had been made an offer by a nurseryman, who wanted to use the land for raising carnations, a development that would have doomed the grove. According to his son, Renoir thought the olives 'the most beautiful trees he had ever seen', and couldn't contemplate their destruction. So he bought the land, originally intending to leave it untouched as a kind heritage site. But his wife Aline insisted that they should build a house there. They moved in and began creating a large and elaborate garden in 1908, and his time there contributed to what is called his 'iridescent period'.

The olives were central characters in this. Their apparent ability to create and scatter patches of illumination made them bespoke trees for the Impressionists, who had abandoned palettes developed for studios where the light was static and came from one direction. Renoir saw the trees as a challenge as well as an inspiration, and understood their legerdemain with light. 'The olive tree!' he wrote. 'What a brute! If you realised how much trouble it has caused me! A tree full of colours. Not great at all. How all those little leaves make me sweat! A gust of wind and my tree's tonality changes. The colour isn't on the leaves, but in the spaces between them.' But he loved their capriciousness. He had a wooden studio with a corrugated-iron roof built amongst the trees, like an exotic garden shed. He liked to pose his nude models on the grass outside, with the sun filtering through the silvery olive branches and dappling their flesh.

The olives (and the garden) are still there at the house at Cagnes, which is now a museum. It's believed that some of the trees were planted by François I, who wanted something to keep his troops occupied during a truce in an early sixteenth-century war. But some of them have the size and archaic weathering of trees known to be 1,000 years old. Unpruned for more than a century, they are untypically tall. The roots have the pitted texture of pumice stone. All of them will have been through Provence's great frost of 1709, which killed many trees down to ground level. A record of the frost remains in their grain pattern – as it does in their branching. The oldest of Renoir's olives have the tortuous growths of all ancient natural survivors: bosses of scar tissue where limbs have been lost, redundant branches wrapped around and almost fused with the trunk, crooks, dog legs, buttresses. The slow ageing of the wood into something resembling the limestone of the Provençal hills anchors the capriciousness of the foliage.

Old and characterful olives are becoming increasingly scarce in the Mediterranean. Renoir saved his grove from clearance for a market garden. Thirty years later Aldous Huxley reported the widespread destruction of olive orchards because peanut oil was in the ascendant. Now, in addition to being cleared for more productive modern varieties, ancient olive trees are being dug up and transplanted to the vanity gardens of rich northern Europeans. Not many survive the trauma, or tolerate their new homes. Those that do are admired as curiosities, but will never have the bright sun of the south flickering through their leaves.

The olive's wild ancestor, the oleaster, and all its shiny leaved, aromatic, brilliantly flowered companions in the Mediterranean scrublands have, for me, become the exotic vegetation of my life. I don't have the right temperament to be a tropical adventurer, so the Mediterranean flora's serendipitous mix of familiarity and strangeness has been a compensation. And there can't be much doubt that this belt of dazzling species – wild tulips, irises, crocuses, peonies, scented sweeps of lavenders, thymes, marjorams, chromatic shrubberies of broom, spurge, cistus, stretching from the Middle East to Iberia – is one of the most beautiful and fragrant plantscapes on earth. A good proportion of our favourite garden flowers have their wild ancestors here. But for me it has the additional thrill of being a dynamic vegetation. For a long while it was believed this was a 'Ruined Landscape', the degenerate remains of some ancient and majestic forest, lost to human plunder. This was a typical anthropocentric myth. The scrubland – the garrigue, as it is called in France – has been there for millions of years. It is a vegetation which has evolved to cope with the environmental challenges of the Mediterranean zone – fire, drought, thin soils, grazing – and its many thousands of species continue to develop, and to find ways of living on vertical cliffs, inside caves, to regenerate from burned stumps, to flower in February snows and then again in the mists of autumn. Perhaps, out of respect for its grace and provenance and majestic inventiveness, we should call it Classical vegetation.

22

Local Distinctiveness: Cornfield Tulips and Horizontal Flax

IN CRETE IN THE SPRING OF 2010 I watched a farmer picking wild tulips by the armful to take to the market in Chania. They were growing like weeds in a wheat field, and were a species found nowhere else in the wild but one small region in the centre of Crete. *Tulipa doerfleri*'s presence on the island is paradoxical. It's globally rare, but depends for its survival not on isolation in the wild but on cultivation of the soil, which disperses its bulblets. It's confined to Crete, but cannot have been here before the introduction of agriculture 7,000 years ago. The flowers have a sultry, Middle-Eastern glamour. The exteriors of the cowled red petals are dusted with satiny gilt, and inside are fired by a fierce scarlet which flares from a dark beauty spot at the base of the bloom. It is one of 159 plant species endemic to Crete, which make up nearly 10 per cent of the total native flora of 1,735 species.

An endemic is a species which is highly adapted to a circumscribed location, and grows nowhere outside it. Its niche can be as big as an island continent, or as small as a cave. The only qualification is that there are no records of the plant occurring of its own accord in wild haunts elsewhere. Endemism is an ambivalent existential state. It may represent the last rites for a wild species (as it did for Wood's cycad), bedded down in a single vulnerable, vanishing habitat with no long-term future. Or a new stage in its evolution, as it adapts genetically to a new refuge. Endemism is still a condition through which species enter and exit the world, as Charles Darwin found with the finches on the Galapagos Islands which were so crucial in forming his theory of evolution.

Sibthorp and Bauer's *Flora Graeca* expeditions provided early evidence of Crete's endemics. Their party landed at Chania, on the north coast, in April 1786 and immediately set about exploring the mountains and gorges inland. On the walls and surrounding rocks of St John the Hermit monastery they found some of the island's specialities:

> [W]e gathered the Ebony of Crete [*Ebenus creticus*] the white Fleabane [*Conyza candida*] the Immortal of the East [*Gnaphalium orientale*] while the soft cotonny Dictamny [*Origanum dictamnus*] carpetted its sides, among the Rocks we found many other curious Plants which the licentious Goats & the burning Sun had spared us ... A Plant that pleased me above all the rest was *Stahelina arborea* of which we brought off a Tree covered with its Flowers & shining with its silver Leaves.

Without quite realising it the pair had struck a seam of plants unique to Crete.

T. doerfleri was in neither *Flora Graeca* nor *Flora Europaea*, published more than a century later. It may be one of the most recently evolved endemics. It was recognised as a distinct species only late in the twentieth century, having been previously lumped in with the very similar *T. orphanidea*. This species differs chiefly in its chromosomal pairings and grows in Greece and western Turkey. Like *T. doerfleri* it is an 'archaeophyte' – that is, not a true native species but introduced, deliberately or accidentally, by early settlers. Changes in behaviour, for instance developing the habit of mimicking an annual, often happen with species that become entangled with processes of cultivation and are subject to unusual environmental pressures and selection processes. No one knows the authentically wild ancestor of these two tulips (*T. kurdica* from northern Iraq might be a candidate), but one can imagine its adaptable offspring moving north-west with the first farmers. One strain became isolated on a secluded Cretan plateau and, over just a few thousand years, evolved into a distinct species.

The Cretan endemic tulip, *Tulipa doerfleri*, growing as a field weed.

Other Cretan endemics are like dialect variations of a familiar vegetal glossary. The woody goosegrass, *Galium fruticosum*, has long, sticky tendrils emerging from a solid trunk, like a jester's whip with a handle. There is a highly local St John's wort, *Hypericum jovis*, confined to the god Jove's redoubts in high mountain cliffs and gorges. The classic site of the small white-flowered catchfly *Silene antri-jovis* is *inside* his subterranean refuge ('Zeus Cave') in the centre of the island. Cretan ebony, seen by Sibthorp, is probably the island's best known endemic. It's a bushy member of the pea family with pink flowers and silver-haired leaves, which grows in great clusters on limestone outcrops. From a distance it makes roadside cuttings look as if they're covered with flowering heather, lightly dusted with hoar frost.

Crete's hoard of endemics is a legacy of its geographical evolution. It began to progress towards island status more than 10 million years ago, as the Mediterranean Sea alternately rose and receded. Plants Crete had previously shared with the adjoining land masses of Greece and northern Africa were cut off when the land bridges were inundated; and as they adapted to new hermetic ecologies, began their slow divergence into distinct varieties and ultimately new species. Then, about 2 million years ago, Crete was rocked by stupendous earth movements that cleaved it with more than a hundred gorges, mostly running north–south. It was as if an already isolated island had acquired a sub-family of subterranean islands, each with its own unique environmental conditions. They became hothouses for the forging of new species.

The deepest and most magnificent of the gorges is Samaria. On the map it seems undistinguished and far from solitudinous, a slightly curved line, which could very well be a road, running north–south in the western corner of the island and ending up in a small resort on the Mediterranean coast. From the high vantage point of the map maker there isn't any hint that it takes you on a climatic journey between the alpine and the subtropical, dropping 4,000 feet in less than fourteen miles. Or that despite being an abyss, an emptiness, it is as paradoxically full of life and disorientating ecological spaces as a forest, including a society of plants which are growing horizontally.

To walk down the Samaria gorge is to feel an echo in your body of the stresses plants must adapt to as they move out of the common greenery of mountain woodland to the parched verticals of the cliff faces. The local writer Alibertis Antonis has described the walk in Delphic terms, as if the journey of the self and the endurance of plants were the same kind of mortal experience. He asks:

> What is Samaria? Is it a bottomless gorge? Is it a perilous but protective path? ... Yes [but] it is also the forest's sensation, the animals and plants' uniqueness, the waters and the springs' charm, the wind's fresh puff rushing through the 'gates'. It is a combination of history and legend. It is the past and the present. It is the fatigue and the audacity. It is the diffusion and completeness.

(The 'gates' – *portales* – are the points where the sides of the gorge narrow to about ten feet. They may have had physical gates once. Eleven people were drowned here in the 1990s when flash floods barged through the gap.)

The track begins high on the Omalos plateau, where there are fields full of more wild tulips, including the endemic pink-flowered *Tulipa cretica*. It's a deceptively cosy start. You open a wooden gate from the road, as if you are about to step into a garden. Aubrietias on cottage walls by the roadside blend seamlessly with wild colonies on the rimrocks of the gorge. There are wooden steps and a handrail. Soon the path steepens, the steps are fewer, and you find you are spending too much time looking at your feet. There is no obvious sightline deep amongst the tall pines, but you're quickly conscious of the kind of space you are entering. It is a world of insistent, precipitous verticals. The roots of ancient cypresses on the slopes above flow down the cliffsides like lava. Huge boulders are scattered about, hurled to the gorge floor by tectonic spasms and mudslides. A pine tree drips resinous candy-floss onto the path as solitary bees drill into its branches. Long trains of nose-to-tail processionary caterpillars climb the trunks in the opposite direction. Pollen falling from the newly opened oriental plane flowers forms gold halos round the rock pools, brief interludes of horizontality.

As the trees open out into patches of sunlight, the flowers begin:

Cretan cyclamen, turban buttercups, and then – longed for but laughably unexpected when I find it while answering 'the call of nature' – the endemic peony *Paeonia clusii*, pure white, gold stamened, and itself crouched under a bush. Its scent is powerful, drowsy, tinged with spice. For a moment it reminds me of the aroma of the moonflower (p. 320), and I wonder if some twist of convergent evolution has made these two species from different continents and different families share a similar perfume, or whether it's my intoxicated memories that are converging.

Halfway down, the immense gorge-side cliffs fill the sky. Their special botanical denizens are known as 'chasmophytes', species adapted to living on arid and almost soilless vertical rock faces. To succeed here they must have root systems capable of penetrating tiny crevices and foliage tolerant of dehydration and fierce sun. It also helps if they have showy flowers to attract insects and make abundant seed. Even quite low down on the cliffs there are exuberant purple-flowered and silver-haired knapweed relatives (including Sibthorp's *Stahelina arborea*), bushes of coronilla hung with globes of cream, the bizarre blue-flowered Cretan wall lettuce, *Petromarula pinnata*, with tall spikes of flowers that resemble kinked propellers, and the best of the gorge endemics, the bushy, bright yellow flax, *Linum arboreum* growing at the very highest levels. Competition amongst these plants isn't often a matter of outgrowing or smothering other species. Many are growing singly or in small groups, surrounded by barren rock in which there are no opportunities for colonisation. Their future depends on chance, the successful lodging of one of their seeds in an untenanted crevice. If it germinates, its life will be precarious, at the mercy of landslides and heavy rains. If it succumbs, its niche may be untenanted for years. The severity of the cliff faces opens up real, if testing, opportunities for the development of variation, and the next seed of the parent species to find a roothold here – perhaps decades in the future – may be subtly different genetically. Sometimes these new variants, given their chance in the isolated nurseries on the cliffs, find their way out into the rocky landscape beyond the gorge.

Down by the *portales*, where the track narrows to a few yards, the cliffs rise an almost sheer 1,000 feet above you. In early summer the wind can become compressed in the thin tunnels through the rock and create vortices which tear up whole plants and whirl them down to the bottom of the gorge like giant thistledown. Yet there is palpable sense of centrifugal force here, of the gorge walls exerting a kind of lateral gravity. It seems to influence the plants, too. You cannot see this from ground level, but many of the chasmophytes are growing horizontally out of the gorge walls. John Fielding's photographs for the definitive *Flowers of Crete* include some taken from perches on the cliff, and show plants – flax, shrubby dianthus, silver knapweed – jutting out from the rock in such defiance of actual gravity and any orthodox leafy attraction to light that you could easily think the pictures have been mistakenly rotated by ninety degrees. Why should this be so? Is it a reflection, an *ob*flection of the powerful penetration their roots must make into the narrow crevices? A way of giving seeds the chance of a wider scatter? Or a consequence of some deep recalibration of the flow of the growth-controlling hormone auxin, alerted that the direction of progress here is sideways?

I think of the way the peculiarities of Samaria are echoed in Crete's other gorges and beyond, in the labyrinths of hundreds of other Aegean islands, and imagine this whole area as a crucible for the evolution of new plant forms, multiplying in its small cracks in space and time.

THE VICTORIAN PLANT THEATRE

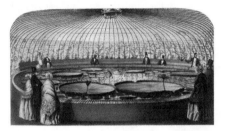

PREDICTED BY NEWTON'S PRISM and the sealed jar of mint in which Joseph Priestley began to uncover the mysteries of photosynthesis, glass became the transforming medium of Victorian life, and especially of botany. It had properties which seemed custom made for the mood of the times. It could foster inquisitiveness and acquisitiveness, encourage aspiration and deep conservatism. Glass could make an enclosure and a window. It enabled private collection and philanthropic sharing, dissolving the barriers between capitalist adventure and communal experience. Its transparency suggested that in an era of such colonial discovery and scientific novelty, there was nothing that could remain unknown. When it came to plants, glass could be both protective and stimulating, excluding cold and the ravages of pests, but always letting in *light*, the Victorians' pervasive metaphor for spiritual and earthly knowledge. More immediately, it could set up a show, even a spectacle. What became known as the Wardian case was the prototype of what were to develop into immense botanical theatres, the nineteenth century's glass arks.

Nathaniel Ward was a doctor working in Whitechapel, one of the poorest and most rundown areas of east London. Professionally, he was concerned about the pervasively grim and devitalised atmosphere of the metropolis, and the impact of its polluted air on his patients' health. He was also an amateur entomologist, and in the autumn of 1829 he sealed a hawkmoth chrysalis in a glass jar, to see if it could successfully over-winter. In the following spring, the hawkmoth hadn't hatched, but in the jar were two lively organisms that had seemingly not been there before. A grass seedling and a small sprig of male fern had sprouted from some moist soil he'd accidentally left behind with the pupa. He had given them no water or any kind of attention over the winter months. He'd witnessed what Priestley had already discovered, that plants enclosed in airtight glass cases are self-sustaining. In sunlight the leaves transpire water, which condenses on the glass and drips down to be absorbed again by the roots. They give off oxygen during the day and reabsorb it at night. It is a closed system, limited only by the amount of nutrients in the soil.

It is odd that Ward was so surprised by what happened in his jar. Priestley's discovery of the elements of gas transfer and photosynthesis in leaves seemed to have become stranded in the confines of chemistry, and didn't influence botanical thought for another half-century. There had even been experiments with miniature greenhouses before Ward's, which had sunk into obscurity. Around 1825 the Scottish botanist A. A. Maconochie wondered if plants accustomed to shade might thrive in a confined, moist atmosphere. He experimented with exotic ferns and club mosses obtained from the Glasgow Botanic Gardens, which he planted out in peat in a goldfish bowl. This was so successful he had a glass case specially built, in which he grew not just ferns but cacti and orchids. But he didn't make his experiment public until 1839, by which time Ward was on the scene.

For years, Ward had been trying to grow ferns in his garden, but found it all but impossible in the toxic atmosphere of the East End. Now, inspired by his accidental experiment, he tried bringing them indoors and growing them in sealed glass jars on his window ledge, and within three years had raised thirty different species, including the delicate filmy fern, which he had collected himself from its *locus classicus* near Tunbridge Wells. He shared his experiences with George Loddiges, one of two brothers who owned a highly successful nursery in nearby Hackney. Loddiges quickly saw the commercial potential of the cases, and made up some stouter and more professional versions. When the perspicacious and influential botanist John Loudon looked them over in

1834, he told readers of the *Gardener's Magazine* in March that Ward's glass cases made up

> the most extraordinary city garden we have ever beheld ... The success attending Mr Ward's experiments opens up extensive views as to their application in transporting plants from one country to another; in preserving plants in rooms, or in towns; and in forming miniature gardens or conservatories ... as substitutes for bad views, or no views at all.

Ward, the public health reformer, had reached the same conclusion and imagined how 'the air of London, when freed from adventitious matter, is as fitted to support vegetable life as the air of the country'. But Ward, the ambitious botanist, had also grasped opportunities hinted at by Loudon. In the book he eventually wrote about his experiments, *On the Growth of Plants in Closely Glazed Cases* (1842), he reflected on the reasons why it was so difficult to convey plants on long sea journeys. He thought it was chiefly due to a 'deficiency or redundancy of water, from the spray of the sea, or from want of light in protecting from the spray; it was, of course, evident that my new method offered a ready means of obviating these difficulties'. Later in 1834 he was ready for a field trial of his invention, and what it might contribute to the international plant trade. Two large cases, filled like botanical arks with English flowers and ferns, were loaded onto a clipper bound for the Antipodes. After a six-month voyage the cases were taken ashore in Sydney and unsealed. The contents were in perfect condition, with one primrose triumphantly in flower. For the return journey Ward had stipulated that the cases should be stocked with native Australian plants, and especially those with a record of being bad travellers. They included a specimen of gleichenia, a small creeping fern which had never survived attempts to ship it to England. The plants were subjected to the severest trials by sea and weather, enduring a storm-racked journey round the Horn, an equatorial crossing and temperatures which varied from −6 to +49 °C (21 to 120 °F). But when Ward went to collect them from the quayside, he found 'the plants were all in beautiful health, and had grown to the full height of the case, the leaves pressing against the glass'.

Soon the cases were in mass production by the Loddiges brothers and others, and in the 1840s, were used to transport 20,000 tea plants from Shanghai to the Himalayas for the East India Company. Later, rubber trees grown from seed gathered in (and some say smuggled from) Brazil were despatched from Kew Gardens to distribution centres across the Far East. Wardian cases became vital tools in the late stages of botanical colonisation.

⁂

The possibilities of what could be done inside glass enclosures reverberated throughout Victorian society. There were dreams of glass-covered shopping arcades, and of an underground railway system mantled by glass (the Crystal Way). In the botanical world, Ward's successful championing of the cause of tax-free glass resulted in a national efflorescence of large versions of his original cases. A suburban hothouse was now an economic possibility, and displays of exotic vegetation that had previously been the prerogative of the rich became accessible to any householder. The window ledge display and the conservatory had been born. The Empire's luxuriant outposts lay just a step beyond the drawing room.

The glasshouse, full of living riches, was an all-embracing projection of the Victorian imagination. In the biggest, the trophies of Empire were kept safely captive but available for all to see. There was a sense of maternal gathering-in, of the glasshouse overriding the constraints of place and time. Rubber plants from South America could grow next to rhododendrons from India. Antipodean proteas might be in flower at the same moment as European apples were in fruit. The mythic idea of the Garden of Eden's 'constant spring' was becoming a reality in the glass gardens of the English shires.

The grandest of all was at Chatsworth, the Duke of Devonshire's seat in Derbyshire, where Joseph Paxton was head gardener and hothouse architect. The Great Conservatory he had designed for the Duke was completed, and ready for its first plants, in 1839. No glass structure quite like it had ever been built before. It was 277 feet long, 123 feet wide and 67 feet at its highest point. It had a wrought-iron viewing gallery, and the aisles were wide enough for three carriages to be driven along abreast. One carriage passenger in 1851 was the King of Saxony, who described it simply as 'a tropical scene with a glass sky'. What an ordinary visitor (there were 50,000 a year) might see was described by Paxton's biographer, Kate Colquhoun:

> Gleaming rock crystals from the Duke's collection were also brought here for display, exotic birds flew among the branches and silver fish swam in pools beneath a plant collection that was simply unrivalled. There were massive, exotic foliage plants and ferns brought from the jungles and mountains of distant continents, orange trees brought from Malta, altingias and araucarias, date palms from the Tankerville collection, the feathery cocoa palm and the giant palm, *Sabal blackburniana*. There were hibiscus, bougainvillea, bananas, begonias, cassias, pepper and cinnamon trees, massive strelitzias – the bird of paradise flower – and hanging baskets of maidenhair fern ... sugar cane, arum lilies and cycas.

It was a theatre made for star attractions, and they would not be long in arriving.

23

'Vegetable jewellery': The Fern Craze

THE FIRST INHABITANTS of domestic Wardian cases, though, were home grown, and more subtle in their appeal. The eruption of interest in fern collecting – which eventually became a 'pteridomania' (from *Pteridus*, the Latin name for bracken) that swept as far as Australia – is one of the more intriguing puzzles of nineteenth-century plant culture. The Victorian mindset – a case study in the tension between contradictions – was enthralled by the sensational and the exotic, sometimes to the point of prurience. But this predilection was balanced, almost penitentially, by a taste for the delicate and the faery, the world of shadow. (David Elliston Allen astutely notes that 'the fern craze opened as men's clothes, quite suddenly, turned black'.) Ferns' intricate, fractal forms – the 'extraordinary exactitude of their ramifications' as one author put it – made them ideal subjects for the contemporary taste for the Gothic. The popular natural history writer Edward Newman made the link explicit: 'The Gothic windows of an old abbey afford many convenient crevices for a pretty fern.' Thomas Hardy, in his descriptions of heathland in *The Return of the Native*, brought a working architect's eye to his description of ferns' intrinsic patterning: 'The ferny vegetation around … though so abundant, was quite uniform: it was a grove of machine-made foliage, a world of green triangles with saw edges, and not a single flower.' They also helped solve a perennial problem for middle-class Victorian men: what do with their womenfolk, an active new generation with time on their hands and open to dangerous temptation. Sending them hunting for ferns – pliant, demure, not obviously sexual creatures which lived in cool, cosseted nooks – seemed a perfect solution. The pastime was 'particularly suited to ladies; there is no cruelty in the pursuit, the subjects are so brightly clean, so ornamental to the boudoir'. It was this audience that the florid garden journalist Shirley Hibberd had in mind for his best-selling book *The Fern Garden* (1869). He enthused about 'vegetable jewellery' and the 'plumy green pets glistening with health and beadings of warm dew' in their glass cells.

But Wardian cases were expensive, ranging from thirty shillings to two pounds, and initially the fern cult was largely confined to wealthy garden enthusiasts and botanists who felt able to manage with a bell jar (a jar big enough to cover seven or eight plants cost a couple of shillings). Pteridomania didn't become a full-blown craze until the 1850s. The repeal of the iniquitous glass tax in 1845 was an essential prelude, greatly reducing the cost of all glass containers as well as windows. The pivotal year was 1851. Ward's original bottle, with the plants still inside and celebrating twenty-one years without watering, was put on show at the Great Exhibition in London. The same year saw the publication of what proved to be the breakthrough popular handbook, Thomas Moore's *A Popular History of the British Ferns*. This was illustrated in colour by W. H. Fitch, one of Britain's foremost botanical artists, and contained a list of places where the choicest and less common species were to be found. Moore's book put into popular currency a trend that had already become evident amongst more scientifically inclined fern lovers: a taste for rare, unusual and frankly peculiar forms. Ferns, as a tribe, seem especially prone to throw up odd forms and 'sports', the fifty or so wild British species being further divided into hundreds of varieties whose fronds are forked, crisped, crested or in some other way marginally aberrant. M. C. Cooke's *A Fern Book for Everybody* (1867; the back cover carries an advertisement for Dick Radclyffe's 'Window Garden Requisites' shop in High Holborn) lists some eighty-five forms of hart's tongue, a common and attractive species whose tufts of glossy, bright green blades grow on rocks, walls, damp hedge banks, and even as epiphytes on mossy branches. It's named from its shape, like a protruding tongue. Local names aren't fussy about which

A Wardian case, full of ferns, suspended in a bay window. From *Cassell's Home Guide*, late nineteenth century.

animal it belongs to: it's also sheep's, fox's, horse's tongue – though the varieties would suggest some alarming glossal pathologies. They include var. *cristatum*, in which the frond divides near the top, with 'each division again and again subdivided so as to form a bushy tuft'; a 'proliferous variety (*viviparum*) [which] bears on the surface of the fronds numerous little miniature plants, which continue adhering to the skeleton as the frond decays'; *laceratum*, described as 'endive-leaved'; and *rugosum*, 'a strange-looking stunted form, with short ragged-edged fronds'. Cooke's guide, written twenty years after the craze reached its height, also sounds a warning about the impact of collecting. The Tunbridge filmy fern, a fragile species with ethereally thin fronds, had been one of the first species to be housed in Ward's cases. Forty years on, the 'rapacity of fern collectors,' in Cooke's words, 'has left very little of this interesting fern to flourish at Tunbridge Wells, on the old station from whence its name is derived'.

Professional fern traders were the real problem. In a manner reminiscent of orchid hunters in the tropics they cleared whole areas, digging up ferns indiscriminately and leaving it to the market to sort out what was interesting. The largest and most elegant native species, the Royal fern, *Osmunda regalis*, was all but eliminated from the Isle of Man. The same thing happened in Cornwall, where one fern seller boasted of having despatched to London a truckload of roots weighing five tons. There were reports of ferns being poached from private estates in Devon and sent up by hamper to Covent Garden. But the amateurs weren't blameless. The 'proper tools' they recommended were semi-industrial: a steel pick, spades and trowels, some lengths of cord and a stout canvas bag. Shirley Hibberd, in another influential book (*Rustic Adornments for Homes of Taste*, 1856), confessed to using a large carpet bag 'fitted with tin boxes, into which the Ferns were laid along with as much of their natural soil as possible' – a rather confined cage for organisms he went on to describe as 'shy wood-sprites'.

The collectors' preferred trophies made for an ambivalent posy. Mostly they were aberrations, mutations, evolutionary dead ends resulting from chromosomal turbulence; and Hibberd's description of ferns as

A fern-decked dress designed by Nicholas Chevalier for Lady Barkly, wife of the Governor of Victoria, 1860. (Pteridomania was as fashionable in Australia as in Britain at this time.)

'plumy ... pets' usefully links them with the freakish, inbred animals beloved in some quarters of the pedigree dog and cat world. Even the most botanically minded pteridomaniac showed no interest in how they had come to be, and what, if anything, they might contribute to the biological future of the fern tribe (the craze peaked just before Darwin's *Origin of Species*). And in the end, like all fashions with no intellectual base or driving force, it became as hypertrophied as a proliferous hart's tongue ('continuing to adhere to the skeleton as the frond decays') and bored itself to death with its own excesses. The most desirable varieties became scarcer and more bizarre. As the originals declined in the wild (some rarer species became extinct in a few locations) virtual ferns sprouted everywhere – in the design of furniture, wallpaper, needle-work, and as shadowy facsimiles made by rendering their images on oiled paper with the soot from a candle flame. The glass ferneries them-selves became rococo fantasies, made to resemble Turkish mosques and European cathedrals. The 'Tintern Abbey Case' was a favourite, with fern-stacked sides and a back panel carrying a facsimile stone wall and stained-glass window, draped with real ivy.

The fern cult says much about Victorian society, its acquisitiveness and readiness to use plants as mirrors of mood and status. In the end the ferns were no more than cultural goods; and if they were an early example of natural capital, they went through the full economic cycle of boom and bust.

24

'The Queen of Lilies':
Victoria amazonica

IT WAS A MEASURE of Britain's dominance of European botany in the mid nineteenth century that its scientific establishment was able to blank out four separate 'discoveries' by French and German explorers of the plant that became the sensation of the Victorian era, and claim it as a regally blessed national treasure. The Amazonian water lily's story is muddied with misidentifications, lost specimens, internecine botanical squabbles and political spin. This makes it hard to trace with any certainty in what order the key events occurred, but highlights what crucial nationalistic tokens plants could become in the nineteenth century.

The first European to see the fabled lily seems to have been Thaddaeus Haenke, a Bohemian scientist employed by the Spanish government to search for plants in South America. In 1801, exploring a marshy canal connected to the river Mamoré in Bolivia, he encountered a water lily of prodigious proportions. He made a few sketches, which clearly identify the species, but no written notes. Nothing would have been known about his sighting if, thirty years later, the French botanist Alcide d'Orbigny hadn't met Haenke's companion on that canoe trip. Father La Cueva told Orbigny that the spectacle of the lily had caused Haenke to be 'transported by admiration', and drop to his knees in praise of the 'author of so magnificent a creation'.

Another Frenchman, Aimé Bonpland, supervisor of the Empress Josephine's spectacular gardens at Malmaison, now enters the play. On Josephine's death, Bonpland decided to move to South America where he'd carried out his early scientific work with the great Alexander von Humboldt. He was exploring rivers near the town of Corrientes in Argentina in 1820, when he saw 'a magnificent aquatic plant, known to the Spanish as *maíz de agua* [water corn]. I described it, and placed it in the genus *Nymphaea.*' The locals ground its roots as flour and Bonpland's interest seems to have been partly in the possibilities of commercial cultivation. But he got trapped by a civil war. The seeds he eventually sent to the Jardin des Plantes in Paris never germinated, and in any case he'd seen a different species from Haenke.

D'Orbigny, who'd seen both the Corrientes and the Mamoré water lilies, was the first to realise they were separate species. The former had leaves which were green on both sides. The latter's were green above and red underneath. D'Orbigny was also the first to put on record the water lily's much older indigenous name, *yrupé*, from the Guarani *y*, water, and *rupé*, a large platter or lid.

A decade later, a young scientist from Leipzig, Eduard Pöppig, became the third European to traverse the entire length of the Amazon, from its source in the Andes to the Atlantic. Sometime in the autumn of 1831 he was exploring one of the river's *igaripés*, side channels, when he came upon a large area completely covered with huge water lilies, 'the most magnificent plant of its tribe,' he declared. He was reminded of the giant Asian water lily, *Euryale ferox*, the Gorgon plant, and thought that this new discovery was from the same family. He accordingly dubbed it *Euryale amazonica*, and preserved parts of the plant in alcohol to take back to Leipzig. He arrived home in October 1832, and a month later published a full description, Latin name included, of the 'new' water lily in a German scientific journal. The news did not cross the Channel, and it was a British-sponsored scientist who, five years later, finally propelled the Amazon water lily into stardom. Robert (eventually Sir Robert) Schomburgk was a Silesian, but his South American expeditions were financed by the Royal Geographical Society. He wrote about what he'd seen in a letter to the Royal Botanical Society, but its contents got their first public airing, according to Wilfrid Seawen Blunt, in front of a meeting of the 'Society of Practical Botanists' in the Strand in September 1837. It sounds like an archetypal Victorian plant salon,

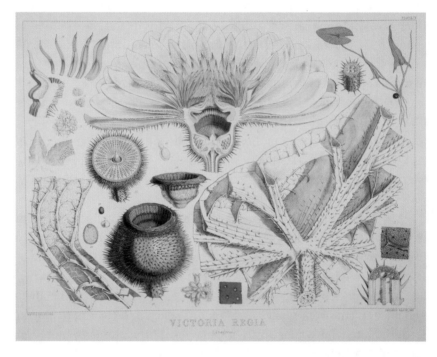

Walter Fitch's anatomical drawings of *Victoria amazonica*, from the specimen growing at Syon House, 1851.

a gathering of artisan enthusiasts and scholarly gentlemen in London's intellectual heartland whom Schomburgk rose to address:

> It was on the 1st of January this year, while contending with the difficulties that nature opposed in different forms to our progress up the River Berbice [in British Guiana], that we arrived at a point where the river expanded and formed a currentless basin. Some object on the southern extremity of this basin attracted my attention. It was impossible to form any idea of what it could be, and animating the crew to increase the rate of their paddling, shortly afterwards we were opposite the object which had raised my curiosity. A vegetable wonder! All calamities were forgotten. I felt as a botanist, and felt myself rewarded. A gigantic leaf, five to six feet in diameter, salver shaped, with a broad rim, lighter green above and vivid crimson below, resting upon water. Quite in character with the wonderful leaf, was the luxuriant flower, consisting of many hundred petals, passing in alternate tints, from pure white to rose and pink. The smooth water was covered with them, and I rowed from one to the other, and observed always something new to admire.

Schomburgk, an ambitious Romantic, wanted to honour the new Queen and suggested the water lily was regal enough to be named *Nymphaea victoria*. The young Victoria, who had only just taken the throne, asked for Schomburgk's drawings to be sent to the Palace, and quickly agreed that her name could be attached to the flower. Almost immediately Victoria's water lily became a symbol of the new monarch and Britain's latest South American colony.

The war of spin that followed degenerated into botanical French farce, though the bickering was taken very seriously by those involved. John Lindley, one of the country's most influential botanists, noticed on examining Schomburgk's dried specimens that the plant was not a *Nymphaea* at all, and in a clever diplomatic move suggested it was from a whole new genus, which was open for naming after the teenage Queen. So the Mamoré water lily, he pronounced in a paper in February 1838, should now to be called *Victoria regia*. But another name – *Victoria regina* – had already been coined by John Edward Gray, President of the Botanical Society of the British Isles, in a paper published five months before. For years the two men squabbled about which name had the prior claim until, in 1847, a German museum curator, Johann Klotzsch, reminded the warring Britons that the plant had already been named in print fifteen years before, as *Euryale amazonica*. Lindley and Gray, insular men for all their botanical nous, knew nothing of this. The 'Queen of Lilies' wasn't a *Euryale* (an exclusively Asian genus), so that family tag had to be ruled out, and they reluctantly agreed that by the rules of botanical nomenclature the first successful bid for each half of the name should stand. So in 1851 the plant was officially dubbed *Victoria amazonica*. The Queen's name had been upheld, but attached to an epithet which linked her, in the mind of a nervous Establishment, with a legendary tribe of marauding female savages. The Director of the Royal Botanic Gardens at Kew, Sir William Hooker, judged that the suffix was 'well enough matched with one of the Furies, but totally unsuited to be in connection with the name of Her Most Gracious Majesty, whom it is intended to commemorate'. Courtly etiquette triumphed over science, and during the Queen's life the name *V. amazonica* was suppressed, and the water lily was always referred to as 'Victoria regia'. The final whiff of powder came from France. Alcide d'Orbigny, miffed at the way his vague description of the other species of *Victoria* had been ignored in Britain, decided – in what has the distinct look of a republican snub to the royalist shenanigans over the Channel – to link the Queen's name with Santa Cruz, the Mexican leader of the Bolivian revolution, when naming his discovery *V. cruziana*.

This comedy of taxonomic manners was trumped by the drama of coaxing the Queen's water lily to flower in her home country, something seen as a political imperative. Schomburgk's first batch of seed had proved infertile. A second batch, purchased from the explorer Thomas Bridges, did germinate, but the two seedlings perished in the 'dark and cheerless' days of winter. In some desperation Hooker instructed two Englishmen living in British Guiana to obtain seed, and send it to Kew in phials of spring water. They arrived safely in February 1849, and Hooker, hedging his bets, divided them between Chatsworth, Kew and the Duke of Northumberland's garden at Syon. At Chatsworth, Joseph Paxton lovingly nursed the seeds in a specially constructed tank kept at 29.5 °C (85 °F). He added a small water wheel to ensure that the lily was 'deluded into thinking she was on her native waters'. Each of the gardens planted out the seeds in conditions they thought most likely to produce the first flower, and the contest turned into a race 'as exciting in its day,' Wilfrid Blunt wrote, 'as Scott's and Amundsen's to the Pole, or the Americans and Russians to the moon'. The plants at Kew and Syon proved incorrigibly shy, but the one in the Great Conservatory at Chatsworth grew so aggressively that its tank had to be enlarged. By the end of September one leaf had a circumference of eleven feet. And on 2 November 1849, Joseph Paxton was able to write excitedly to the Duke (then in Ireland): 'Victoria has shown a flower!! An enormous bud like a poppy head made its appearance yesterday. It looks like a large peach placed in a cup. No words can describe the grandeur and beauty of the plant.' The Duke rushed home, and had a flower and a leaf sent to the Queen at Windsor. The historian Isobel Armstrong (though she wrongly attributes the name *V. regia* to Paxton, who she imagined coined it in an act of loyal fawning) describes 'the lily enterprise' as 'an aristocratic stunt, the deliberate creation of a myth of Brobdignagian botanical birthing is some ways akin to sensational events – the balloon flight, animal freakery – arranged by popular urban gardens'.

Perhaps so, but one of these urban gardens was Kew, and its metropolitan position and loyal fan base ensured that the lily's flowering was far from an elitist 'stunt'. By the late 1840s Kew was being visited by up to 10,000 people daily. Admission was free, and the deliberate policy of the director, Hooker – high minded but democratic – was to tempt the 'frequenters of pot-houses and skittle-alleys in "London's dirty, crowded pestiferous courts" to pass their morning in Kew's "perfect paradise"'. The water lily bloomed the following summer, and continued its display right up until Christmas. It proved one of Kew's greatest attractions to the plant-infatuated citizens of the capital, and Hooker had a much larger tank built to show it off to its best advantage. When it was first exhibited, thousands made the trip to Richmond to experience the full sensuality of its flowering. Kew's specimens were punctual timekeepers.

At about two o'clock each afternoon (the garden fortunately opened to the public at 1 p.m.) the new white buds, the size of tennis balls, began to emit a strong aroma, variously compared to melon, strawberry and pineapple. A few hours later the petals opened and began to change colour to rose pink. Towards ten in the evening they started to close. The flowers' slow decline continued the following day, when the fading petals became 'a drapery of Tyrian purple' until they finally sank beneath the water. Alas, as so often at this time, European horticultural reflexes proved quite inappropriate to the raising of tropical plants, and *Victoria*, deprived of decent ventilation and flexible heating, had to be moved again.

I've yet to witness the full flowering of *Victoria*. Years ago I kept a daily vigil by the specimen in the tropical biome at the Eden Project in Cornwall, but the flowers were always wide open or tight budded, and I never got to see – or smell – its baroque unfurling. It was some consolation that the legendary Amazonian botanist and onetime director of Kew, Sir Ghillean Prance, was also in the biome on one occasion. He told me how he had witnessed the rite himself in the wild and explained what the extravagant flowering meant for the plant itself. The opening flower buds (in a 'male' state at this point) raise their temperature until it is eleven degrees Centigrade higher than the air outside, to add the lure of warmth to the entrancement of smell for the pollinating beetles. Once the insects are inside, the flower shuts, trapping them for twenty-four hours, during which they pick up a mass of pollen on their bodies. The next day the flower opens again, now with the female parts prominent, and beetles make a rapid escape, hopefully flying to another open flower to pollinate it. Meanwhile, beetles that had been trapped elsewhere carry their load of pollen to the fading 'female' flowers.

My chance to see the rite of opening, though not of pollination, came when the Amazonian water lily at Cambridge (the nearest botanic garden to my home) began a run of prolific flowering in the late summer of 2013. I got permission to visit the glasshouses after hours, and a promise of an alert call when a bud began to show signs of movement. The Cambridge specimen is *V. cruziana*, which is hardier in cultivation than its cousin

V. amazonica, but shares with it a mischievousness during flowering. Sometimes a likely bud stays teasingly tight; sometimes it opens in an unpredictable rush. When the call came that a flower had opened unexpectedly in the night and another was due that evening, I sped to Cambridge and was escorted to the promising debutant by Alex Summers, the Glasshouse Supervisor. It did seem to be swelling infinitesimally, and there was the faintest smell of pineapple around the pool. But, as they say, a watched kettle never boils, and the bud remained stubbornly cool and closed. By ten o'clock we'd abandoned hope.

But the flowerings here, even when unwitnessed, aren't futile. In the absence (so far) of tropical scarab beetles in Cambridgeshire, Alex had to stand in as pollinator. Up to his thighs in the outside lily pool, he began pushing cheap paintbrushes into the golden stamens of the previous night's voluptuous flower. The scene was like a séance: the small circle of expectant viewers, the shadowy backcloth of exotic trees reflected in the glass, the lily lit up by the ectoplasmic green glow from half a dozen smartphones. 'Channel the beetles, Alex,' a spectator urged. Well, Alex did, very successfully. A great quantity of pollen was gathered, and transferred the next day to the flower which had opened in its own good time after we'd all gone to bed. A month later I heard from Alex that his beetling had midwifed 194 seeds 'each similar in size to a ball bearing'.

In the end it was the leaves more than the flowers that were to prove the enduring wonder of *Victoria amazonica*. Richard Spruce, another Amazonian explorer, seemed prescient in 1849 (the year the lily flowered at Chatsworth) when he gave a thoroughly industrial account of the water lily's architecture:

> The impression the plant gave me, when viewed from the bank above, was that of a number of green tea trays floating, with here and there a bouquet protruding between them; but when more closely surveyed the leaves excited the utmost admiration, from their immensity and perfect

symmetry. A leaf, turned up, suggests some strange fabric of cast iron, just taken from the furnace, its ruddy colour, and the enormous ribs with which it is strengthened, increasing the similarity.

The ribs radiate from the centre like the spokes of a wheel, flatten as they extend outwards, and then split up into as many as five forking branchlets, so that at the edge a single rib may have separated into thirty thinner struts. This forking is the key to the leaf's stability.

Several expensive folios of paintings of the lily's leaves were published, including one by Walter Fitch, whose lithographs were reduced from originals twenty feet across. The leaves began to influence interior decor, too. Many designers were struck by the distinctive shape and structure of the leaves, incorporating them into chandeliers and gas brackets. One marketed a papier-mâché *Victoria* cradle. In November 1849, the Amazonian Indians' custom of bunking down their children on the leaves while they were working was repeated with great success at Chatsworth. The Duke of Devonshire and Lady Newburgh placed Joseph Paxton's seven-year-old daughter Annie, dressed as a fairy, on one of their home-grown leaves. (Paxton had first tested the leaf, whose tissues were delicate enough to be pierced by a straw dropped from a few feet, with a heavy copper lid, as much out of technological curiosity, I suspect, as concern for his daughter's safety.) Douglas Jerrold wrote a short doggerel to commemorate the event ('On unbent leaf in fairy guise,/ Reflected in the water,/ Beloved, admired by hearts and eyes,/ Stands Annie, Paxton's daughter') and on 17 November the *Illustrated London News* published an engraving of Annie afloat. It is a strange and unsettling picture. Annie, far from snoozing like an Amerindian child, is standing stock still, looking shy and awkward. A small scatter of people are leaning on the rails round the lily's enclosure and gazing laconically, *assessing* her. The scene is uncomfortably like an auction, with Annie posed in the role of surrogate Indian. The Victorians were quite capable of loving the plant while despising its origins.

Meanwhile, Annie's father's skill in designing glasshouses had lifted him above his earlier station as a gardener. He had already begun using the ribbing of *Victoria*'s leaves in remodelling the Chatsworth *refugium*

A 1936 photograph showing the extraordinary engineering of *Victoria amazonica*'s leaf, which had inspired the metal framework of the Crystal Palace.

in which it grew. In 1850 he gave a talk to the Royal Society of Arts about his glasshouse, during which he demonstrated, and gave credit to, the structure of the water lily's leaves: 'You will observe that nature was the engineer in this case. If you will examine this, and compare it with the drawings and models, you will see that nature has provided it with longitudinal and transverse girders and supports, on the same principle that I, borrowing from it, have adopted in this building.' Shortly afterwards he won a competition, set up by Royal Commission, to design and bring into being a building to house the Great Exhibition planned for London in 1851. The result was the Crystal Palace. In designing it Paxton used *Victoria*'s device of developing a few main ribs into a series of thinner struttings, bound to one another by many fine cross-ribs. No large building had ever been built like this chimera of steel and cellulose before, but every large glass building since ultimately derives from it.

25

A Sarawakan Stinkbomb:
The Titan Arum

HENRY GOSSE, in *The Romance of Natural History* (1861), linked the exotic wild flower and the tamed house plant in an almost prelapsarian account of his first experience of a tropical forest, describing 'a beauteous gloom – rather a subdued and softened light, like that which prevails in some old pillared cathedral, when the sun's rays struggle through the many-stained glass of a painted window. Choice plants which I had been used to see fostered and tended in pots in our stove houses at home, were there in wild and *riant* luxuriance ...' The Victorians, for all their love of order and conquest, weren't averse to the chaos of tropical vegetation, to *jungle*, to the thrill of monstrosity and what it said about the superiority of their own species and civilisation. In June 1889 Kew was host to a plant in some ways more compellingly memorable than even *Victoria*, and which was described as 'one of the sensations of the London season'. In 1878 the young Italian botanist and traveller Odoardo Beccari had discovered an aroid of huge and scabrous proportions in the Sumatran rainforest. It's best visualised as a magnified and bloated version, with a lurid crimson sheath and disgustingly foetid spike, of the European wild arum (Lords and Ladies, an abridged version of the original name, [the] Lord's and Lady's [genitalia]: we had bawdy plants too). Beccari described how

> The single flower, or more correctly inflorescence, and the tuber (from which it springs almost directly), form together so ponderous a mass, that for the purpose of transporting it, it had to be lashed to a long pole,

the ends of which were placed on the shoulders of two men. To give an idea of the size of this gigantic flower, it is enough to say that a man standing upright can barely reach the top of the spadix with his hand, which occupies the centre of the flower, and that with open arms he can scarcely reach half way round the circumference of the funnel-shaped spathe from the bottom of which the spadix arises.

Beccari was able to take seeds of the aroid, by now admiringly dubbed *Amorphophallus titanum*, back to Italy, where they were grown on in the garden of his friend, the Marchese Corsi Salviati, in Florence. The Marchese presented some of the young plants to Kew, along with an altogether more extraordinary offspring: a life-size painting of the Titan, on a canvas eighteen feet by fifteen, showing a leaf growing out of the ground, and two Sumatrans carrying a tumescent inflorescence lashed to a pole. For a while the picture adorned the roof of the Orangery at Kew until offended officials banished it to a less brazen position. Victorian prudishness has been persistent. A photograph in the Kew archives from 1938 seems to show the picture consigned to the ceiling of a storehouse, and with the arum's swollen flower spike painted out, leaving just the carrying frame. And when David Attenborough filmed the living plant for his BBC series *The Private Life of Plants* in 2008, he declined to use the scientific name *Amorphophallus* because he thought it 'inappropriate' for a family audience.

But no such delicate sensibilities were evident towards the plants themselves in the 1880s. They took ten years to reach the flowering stage, but their growth in the last two was prodigious. In 1887 the biggest tuber was three feet nine inches in circumference, and the leaf six feet tall. By 1889 the 'bud' was growing at the rate of three inches a day. The Titan finally bloomed at 5 p.m. on 21 June, revealing one of the strangest and most suggestive floral structures ever seen in Britain. The central spike, or spadix, is hollow, and resembles, according to one polite description, an immense, fresh-baked baguette. The surrounding spathe is a frilled and deeply furrowed sheath, more than a yard in diameter at its top and lined in burgundy red. During flowering it emits an over-whelming stench, described by one visitor as 'suggesting a mixture of rotten fish and burnt sugar', and its general miasma of decaying meat

Timelapse photos of the emergence of the Titan arum's flower spike, which grows three inches a day.

(one popular name is 'corpse flower') almost certainly evolved to attract pollinating carrion beetles and flesh flies. Echoing *Victoria amazonica*, the temperature of the spadix rises during flowering to almost exactly human-body warmth – again perhaps to hoodwink flesh flies.

Kew's Titan had been well advertised. Enormous crowds gathered to witness its moment of glory, but weren't prepared for the full horror of the 'smell experience'. One of Kew's artists, Matilda Smith, daughter of the director, was given the brief of drawing what had now been nicknamed the 'giant stinker', and was solemnly thanked by the editors of the *Botanical Magazine* for her 'prolonged martyrdom' at the easel. It was not her only trial at the hands of Beccari and his fancy for the malodorous. He had also discovered a new orchid in Sarawak (subsequently named *Bulbophyllum beccarii* in his honour) whose flowers were small and nondescript, but which in his words smelt 'like a thousand dead elephants'. When it bloomed at Kew it 'rendered the tropical Orchid house … unendurable during its flowering'. The hapless Ms Smith was taken ill while trying to paint it, and had to abandon the project.

26

Harlequins and Mimics: The Orchid Troupe

FOR TWENTY HAPPY YEARS I was the custodian of a small beech wood in the Chiltern Hills. It had a fabulous collection of ground plants with subtle and elusive charms: the February-flowering spurge laurel, the year's first hint of honey; the fretwork and white linen of wood anemones in April; the moment in late May when the wash of fading bluebells turned to the colour of wood smoke. We had three species of fugitive orchid, including violet helleborine, the last bloom of the year, whose purple-tinged stems and leaves were almost invisible in the shade of the deepest thickets. But it was the ferns, the woodland floor's plumage, its green peacock tails, that I loved most. There were nine species in the wood's sixteen acres, a beguiling intrusion of Celtic scrolls and soft accents into a region that had always seemed to me the epitome of middle-England flintiness. Shield ferns haunted the wood's medieval banks, softening the boundaries. Scaly male ferns fanned the bluebells with fronds of lime green on deep-orange ribs. And for all the wood's antiquity, they shifted about in it almost capriciously. A colony of the softly serrated lady fern somehow found its way to the deepest darkest hollow, defying all theories about its reluctance to move to new sites. The whole wood, 600 feet up and with its own microclimate, was a Wardian fern case on a landscape scale.

After a long illness I left the Chilterns in 2002 and moved to south Norfolk – a flat land, wet on the ground but dry in the air and stripped of most of its old woods. A bell jar full of ferns from my retreat amongst the beeches would have been the best possible memento of the life and landscape I'd left behind. But I've never had the collecting instinct, especially for living things, and the only botanical trophy that made its way east with me was a painting of an orchid. It wasn't one of our wood's hermits, and not even a British species. But with hindsight it was a good transitional object, free of cloying nostalgia and linking the places I'd migrated between. East Anglian wetlands and the chalk downs and beech woods of the Chilterns don't have much in common as landscapes. But they are both infertile, marginal lairs, classic orchid terrains, and these glamorous flowers, whose lives are so redolent of metamorphosis and of cryptic, subterranean partnerships, had been familiars since I was in my twenties. So were they for the Victorians – though perhaps at a more acquisitive level. But I bought the picture, so have no grounds for self-righteousness.

It was a print (number 32 of 40, and life sized, or so I imagined) of *Paphiopedilum sanderianum,* by the Scottish painter Dame Elizabeth Blackadder. I'd always loved Blackadder's flowers. They're fastidious without being fussy, not rooted in any kind of landscape or even the soil but weightless, airy, as if they have been artfully scattered about the canvas by a breeze. When I saw this print in a gallery, I bought it as much because it was by her as for its orchidaceous subject.

But Sander's *Paphiopedilum* might have evolved especially to indulge her gift for vegetable levitation. She had pictured two flowers, one entirely free floating, the other emerging from a tuft of leaves which vanished under the left edge of the frame. The flowers had the look of Samurai warriors in ceremonial armour, with tall, tapering, striated helmets at the top, and tunic-like pouches, the colour of beaten copper, below. And from a pair of projections just below the 'head' flowed two extraordinary streamers – spiralling, tiger-striped pigtails. On my print they measured ten inches in length. I knew nothing about the real plant, beyond assuming that, in common with most tropical orchids, it was a kind of epiphyte.

In that first year in East Anglia I felt, as I wrote in a memoir of the experience, like a forest epiphyte myself. I was lodging in the Waveney

Valley, in a sixteenth-century farmhouse of the kind usually called 'half-timbered', in which there seemed to be more oak inside than out. My room had oak floorboards and an oak desk. Oak beams, faded to the colour of old bone, ribbed the walls and the corners of the ceiling. Because there was nowhere else to put it, I hung my orchid from one of the vertical beams. And one night, when I'd turned the lights low, the room's hard edges – book spine against lime plaster, flex against floor-board – began to soften, and the *Paphiopedilum* suddenly seemed to be growing *out* of the beam, just as I imagined it did in its native habitat.

I sat in this indoor forest glade most evenings, trying to make sense of the new landscape I'd fetched up in, and found that, without especially meaning to, I was reading increasingly about orchids, *local* orchids – real, fabulous, idiosyncratic, protean. Orchids lost, found, lost again. The fens and wet meadows that edged the Waveney Valley seemed to be an Elysium for them. A Lammas-land pasture stippled with the delicate stained glass of green-veined orchids lay a couple of miles to the north. A variety of bee orchid (var. *chlorantha*) with greenish-white petals as diaphanous as onion skins was almost *appellation contrôlée* here, as locally accented as an artisan cheese. And in a fen just a few hundred yards across the fields from the house, in 1936, the legendary orchid prospector J. E. Lousley (a banker in his day job) had discovered a subspecies of early marsh orchid with 'straw-coloured' blooms, known as *ochroleuca*. It sounded as luscious as clotted cream. Roydon Fen was one of only three sites in Britain, and the flower, so my books said, had not been seen anywhere since 1970. It was presumed extinct – or maybe just gone to ground. It went to ground somewhere in my head, too, part of a map of lost treasure.

If this wasn't enough, a few hundred yards away in another direction was Blooms Nursery, where, in a pavilion resembling Kubla Khan's pleasure dome, all manner of exotic orchids of bewildering shapes and hues, and as perfectly finished as bone-china ornaments, were sold as house plants and last-minute Valentine's Day gifts. These were orchids as commodities, mass produced by hybridisation and cloning. They rarely lived beyond their first flowering. I seemed to be triangulated by orchids, not just by the literal geography of these three places – bedroom wall, valley fen and plant supermarket – but by three kinds of orchid experience. The ethereal image in Elizabeth Blackadder's watercolour seemed to be a dream or distillation of the wilderness, which refracted in one direction to the living, if humbler, orchids of an English meadow, and in another to the merchandised products of the garden centre, also alive in a literal sense, but somehow fossilised, plant tissue reduced to cosmetic token. Each, in its own way, demonstrated the extraordinary power orchids have come to hold over the imagination: a captivation which stretches back at least three centuries, and can't be explained just by the compelling appearances of their flowers.

I eventually learned more about *P. sanderianum*, where it grew and what it looks like in real life, but I can't say this cast much light on what lies behind the orchid family's mesmeric powers. Close to, the copper tunic resembles a small pitcher, or a ballet shoe, and puts the plant in the large group of slipper orchids which includes, as a distant cousin, the European lady's slipper, *Cypripedium calceolus*. The pendulous streamers are, in fact, petals, which can grow up to three feet long and act as a lure for pollinating insects. It was first discovered in a rainforest in Sarawak, Borneo, by the German plant collector J. Förstemann in 1885, growing not, as I'd imagined, as a conventional epiphyte, trapezing from a roothold high up in the canopy, but anchored in the ground, often to a tree root in steep and bony limestone outcrops. It was named in honour of the Victorian orchid entrepreneur Frederick Sander, whose ambition to bring orchids within the reach of ordinary people was achieved by stripping whole areas across the globe clear of their plants. Unsurprisingly, it was never seen in its original site again, and was thought to be extinct in the wild. Sander's slipper became the 'holy grail of the orchid world', and when it was rediscovered in 1978 by the collector Ivan Nielsen, growing near Fire Mountain in a remote region of Sarawak, it sent ripples through the entire orchid world. Professional botanists, commercial growers, obsessive collectors and frankly criminal looters all wanted to get their hands on it, as booty for research, artificial propagation, or high-end trading. Many tried to track down the site, and a few succeeded. By 1989, when trading in wild specimens was made illegal, a cultivated Sander's slipper could fetch up to $3,000.

Sander's slipper orchid, painted by the orchid collector Frederick Sander himself in 1891.

What is intriguing in this tangle of aesthetic enthralment and unscrupulous trophy hunting is the disproportion between the obsessional nature of orchid fancying and the character of the plants. Species such as, say, the Amazonian water lily or the Titan arum, have appearances, histories and life cycles just as glamorous as orchids', but haven't generated the same addictive fascination. In his picaresque enquiry into modern collecting, *Orchid Fever: A Horticultural Tale of Love, Lust and Lunacy*, Eric Hansen describes a world as manic as the seventeenth-century Dutch tulip cult. His stories of smuggling, grand theft and gunfights in the glasshouses smack of the dark frenzy of drug trafficking. Thailand alone exports more than $250 million worth of plants every year, many of them from the wild, and not all of them procured legally. The retail trade in commercially grown orchids in the USA exceeds $150 million, and there are half a million dedicated collectors there. Worldwide, the turnover of the orchid industry is valued at more than $9.5 billion, and the estimated tally of 150,000 artificially bred varieties is increasing by more than 200 a month.

Yet the comparison with what has come to be called 'Tulip Fever' doesn't quite work. Tulips in seventeenth-century Holland were no more than a fashionable currency. Increasingly bizarre and often ephemeral varieties were bred to raise the market ceiling but the identity and character of the plants was irrelevant. They might just as well have been rare dahlias, or filigree bonsai. But there seems to be some quality or *mana* attached uniquely to orchids that underlies their universal appeal. An aura (a *generic* aura: they make up a huge and disparate family of some 25,000 species) of voluptuousness, exoticism, perhaps even decadence; a hint of ambiguity in the seductive beauty, to humans, of flowers designed to lure rampant insects. There is a mundane allusion to sex even in the family name, from the original Greek *orkhis*, meaning testicle, alluding to the shapes of the underground roots. (Orchids were 'ballockworts' in Middle English.)

It was during the nineteenth century that the mysterious glamour of the flowers began to take hold. Amongst the more reverent supplicants was Henry Thoreau, who described the US East Coast purple fringed orchid as 'a delicate belle of the swamp ... A beauty reared in the shade of a convent, who has never strayed beyond the convent bell.' Others relished what they saw as an essentially decadent motif running through the family. The anti-hero of the fin-de-siècle aesthete Joris-Karl Huysman's famous novel *Against Nature* has a morbid taste for *Cypripedium* orchids: 'They resembled a clog, or a small oval bowl, with a human tongue curled back above it, its tendon stretched tight just as one sees tongues drawn in the illustrations to works dealing with diseases of the throat and mouth.' Marcel Proust, in *À la recherche du temps perdu*, uses cattleya orchids as signifiers in the private language of Charles Swann and Odette de Crécy: 'To make Cattleya was shorthand between Swann and Odette for making love, ever since an episode in her carriage when her horse shied at an obstacle and Swann sought permission to restore the *Cattleya* to her bodice' (she was wearing more orchids in her hair, and carried them in a bouquet). H. G. Wells wrote a brisk but gory horror story called 'The Flowering of the Strange Orchid'. A London bachelor called Winter-Wedderburn buys an unnamed rhizome at an orchid show, which he plants in his greenhouse. Later his housekeeper discovers him almost expired, overcome by the sickening miasma of its scent, and with its rapidly extending aerial rootlets anchored in his neck. They reminded her, she reflected later, of 'little white fingers poking out of the brown'. Well into the twentieth century orchids play sinister and often double-entendre roles in crime fiction. At the start of Raymond Chandler's *The Big Sleep* Philip Marlowe meets General Sternwood in the conservatory of his Hollywood mansion, where the air is 'larded with the cloying smell of tropical orchids in bloom'. The General passes on his own views about the plants he nonetheless keeps close to him: 'They are nasty things. Their flesh is too much like the flesh of men. And their perfume has the rotten sweetness of a prostitute.'

These metaphors are cumulative. Out of the chemistry of association and provenance, orchids become more than themselves, even the most humdrum basking in the reflected glamour of the family mantra – *orchid*: the bloom of the exotic romance and the privileged hothouse. Give a dead-nettle, with its symmetrical arrangement of pink labiate flowers, the tag of orchid and it might attract the same reverence. But the provenance of an image, or a reputation, has to start somewhere,

The orchid's perennially voluptuous image. *Cattleya Orchid and Three Hummingbirds* by Martin Johnson Heade, 1871.

and one important ingredient of orchids' appeal is that so many of them – or bits of them – resemble something else, especially elements of human anatomy: diseased tongues, naked limbs, exploratory fingers. The basic template of the orchid flower itself is vaguely animal-like, often hominid: a manikin with a small head (the column) surrounded by petals which, above, often form a hood or headdress, and laterally, represent arms. Below is a lip (the labellum) which can be broad, as in a skirt, or forked, like a pair of legs. Hybrid forms, crossing the bound-aries not just between species but whole classes of being, have always touched the human imagination, as in the mythology of chimera like the vegetable lamb.

Even amongst the limited roster of European orchids, the power of these suggestive resemblances shows up in their popular names. The gardener Philip Miller noted this in 1740, remarking how the flowers sometimes represent 'a naked Man, sometimes a Butter-fly, a Drone, a Pigeon, an Ape, a Lizard, a Parrot, a Fly, and other Things'. There are also bee, spider, bug, woodcock, tongue and lady orchids. There is a military or soldier orchid, named, according to the Elizabethan botanist John Gerard, because it has 'little floures resembling a little man having a helmet on his head, his hands and legges cut off'. And a ghost orchid in ectoplasmic pink, which appears in the darkest woods as irregularly as a poltergeist (and occasionally blooms *under* the ground). The lady orchid's flowers support splendid pink-spotted crinolines. Those of the man orchid are undeniably anthropomorphic, but are spindly, jaundiced and topped by a hood resembling the bulbous green craniums of the Venusian Treens in the Dan Dare comic strip. The hybrid between the man and the rangy-limbed monkey orchid (Miller's 'Ape'), which crops up rarely in southern Europe, has been tagged the 'missing link orchid' by some waggish orchidophiles. Similarly, the common English name of the Mediterranean *Orchis italica* – Italian man orchid – is sometimes derisorily parsed by British botanists as 'Italian-man orchid', because of the shortness of the little pink appendage between its legs; their Italian counterparts responding by stressing that, proportionately, it is almost as long as the thigh.

All this name gaming and pattern recognition has a touch of the

Mexican craze of looking for the face of the Madonna in burnt tortillas. You see what you believe in. But not in the case of the lizard orchid of continental Europe, whose flowers are persuasively reptilian to any viewer. They have long forked tails with a *sanderianum* twist, flexed olive legs and pale green heads. They stink of goat. The most Gothic – but not at all inaccurate – description is by the Belgian symbolist poet Maurice Maeterlinck in 1906:

> It is symmetrically adorned with vicious three-cornered flowers of a greenish white stippled with pale violet. The lowest petal, decorated at its source with bronzed caruncles, with Merovingian moustaches, and with ominous lilac buboes, extends endlessly, crazily, improbably, in the shape of a twirled ribbon, of the colour of a drowned person whose corpse has been in a river for a month.

The lizard orchid has nothing whatever to do with lizards, and most such names can be put down as similes, based on fanciful and light-hearted resemblances. It is only to us humans that the frog orchid suggests an amphibian, and the vague likeness has no biological function at all. But there is one group of orchids whose similarities to other organisms are more than coincidental, and which bring biological mimicry and literary metaphor into testing confrontation. There are some twenty true species in the European genus *Ophrys* (and many in related tropical families) all of which have a superficial resemblance to wasps or bees – the sepals or outer lateral petals extended like wings, the inner petals shrunk to resemble antennae, the body oval and brown. The bee orchid is the best known, but I have to say that even here the resemblance is very crude, and the flowers are less like living insects than toys, stitched together from starched paper and velvet. Some early botanists imagined these likenesses were intended to frighten insects away, so that they would not damage the flowers. But it's now known that this thespian posing is, on the contrary, an attracting device. Its purpose is to persuade passing male insects that the flowers are females of the same species, so that they jump on their backs and engage in what is called, with the stiffest of scientific po faces, 'pseudocopulation'. During this process the insect's head is close to the orchid's sexual parts (real, not pseudo), and with luck picks up pollen-bearing growths (pollinia) which it later deposits on the stigma of the next orchid flower it embraces. I've never been wholly convinced by the popular explanation for this unconventional coupling, with its implicit assumption that a visually acute insect, which would never mistake even a closely related species as a potential mate, would hump a flower on the sole basis of a vague physical resemblance. Insects use many more senses and cues in navigating their world than we do, and I've always suspected other signals, most probably phero-monic scents, must be the primary cues in attracting the males. This is the consensus of modern scientific opinion, and it is now known that the male insect mates with the female pupa in the wild, not the fully hatched insect. This is a more amorphous creature than the mature adult, and may be what the orchid body is mistaken for, once it has been located by scent.

Pseudocopulation is a fact, and has been observed many times. The most graphic account is one of the earliest. Colonel M. J. Godfery published his classic *British Orchidaceae* in 1933. It is a book of extraordinary detail and loving attention – of many kinds. His wife Hilda painted the immaculate illustrations, and she is the only dedicatee of a book (she died before its publication) I have seen commemorated with an 'In Memoriam' photograph. She is posed like a Green Woman against the Godferys' copious shrubbery.

The pair were on a hillside near Chambéry in the French Alps in May 1928 when they witnessed at close quarters the first stages in the pollination of the fly orchid, *Ophrys insectifera*:

> It [the pollinating insect] is hard to see when quiescent on the flower, the closed wings agreeing with the contour of the lip, the gap between the thorax and abdomen seen through the wings giving much the same impression as the leaden oblong marking on the middle of the labellum, and the antennae resembling thread-like petals. It alights on the lip head uppermost, and rests there with quivering wings and waving antennae, doubtless a preliminary phase of courtship, sometimes for three minutes. Its actions made it quite clear that the wasp regarded the lip as a female of its own species.

A century and a half earlier John Langhorne had penned some coy quatrains about this conjoining that are a choice example of inadvertent orchid erotica:

> See on that flow'rets velvet breast
> How close the busy vagrant lies!
> His thin-wrought plume, his downy breast,
> The ambrosial gold that swells his thighs!

Godfery's account is riveting and convincingly precise. But his words still seem to me to overplay the importance of purely visual mimicry, and aren't immune to metaphor. I talked the problem through with my old friend and botanical mentor Bob Gibbons. He was resolute in defence of the visual deceit. Why else would *Ophrys* flowers – all of them – have even the vaguest resemblance to insects? But he wondered if perhaps the *feel* of the flowers' bodies acted as an additional aphrodisiac to the hormonally charged males. It was an agreeable thought: furry bees clasped to velvet midriffs, 'restlessly vibrating', and I had the notion of abandoning my usual siding with Keats's passive flowers – 'budding patiently … taking hints' – and trying to see the transactions from the point of view of the randy and hurrying bee. Fly orchids – the most convincing mimics in the *Ophrys* genus and the subject of Godfery's historic observations – felt like the best bet. I persuaded Bob to post me a couple of spikes from the Corbières in France (where they're common enough for guilt-free picking) so that I could get to intimate grips with them myself.

There was another, more personal reason for choosing fly orchids. Now thoroughly alert to all hints and echoes of mimicry, I remembered that when I'd lived in the Chilterns, I'd been unconsciously insect-like in my ritual spring searches for them, which happened in the heart of my home patch and tested my hunter-gatherer's instinct. They were the most challenging species to find, growing in grass at the edge of chalky beech woods and in the light shade of hazel scrub. Since they were rarely more than a foot in height, and the blooms not much bigger than wasps, they were almost impossible to see amongst the early summer vegetation. I used to stalk them, crouched down like a beetle, scrabbling under hazel boughs, hoping to notice an incongruous dark shape at an odd height in the grass. I had my own pet colony, in a hazel coppice last cut by Italian PoWs in the 1940s. I liked to believe I was the only person who knew of it, and had no qualms about doing a little annual gardening, pulling out clumps of the aggressive dog's mercury that was always threatening to overwhelm the precariously fragile stems. It never made them easier to find the following year, which perhaps made the hunt more deeply satisfying and helped fix it in my memory as embodying the essence of that particular moment of the spring. The vegetation seemed to be an undifferentiated curtain of verticals, the sun always coming from behind me, the faint dusts of bluebells and cow parsley still in the air. What gave the fly orchids away, as often as not, was the thin band of shiny blue that stretches like a belt across the middle of the body. The flowers are dark and narrow, the inner petals curled up into stumpy horns, the lip with slight swellings down each edge, which might or might not resemble folded wings. Later I saw photos of the pollinator, a digger wasp, *Argogorytes mystaceus*. It is wasp-waisted, striped with yellow, and bears only a passing resemblance to a fly orchid bloom. But maybe the female pupa, not yet fully striped, is more suggestively arousing. One photograph I saw showed three male wasps on top of each other on a single bloom, in the most orderly of floral gang-bangs.

The spikes arrive from France by express delivery, packed in a bizarre assortment of damp leaves rife with Bob's botanical in-jokes. The flowers are rather faded versions of the vivid garnet and sapphire blooms I remember from Chiltern springs. Several days in the post and several years of romantic fantasising have made a difference. I cut the two freshest from the stem, and indulge in some pseudo-foreplay. I sniff them, lick them and rub them against my lips. They have no scent whatsoever but I can feel a texture on the body that reminds me of the pimply surface of a tongue. What I do next is in full knowledge that I'm not really qualified to interpret what I see, and that my perceptions bear no relation to an insect's. But I slide them under my stereo microscope anyway.

At 10x the body of the flower is still just about legible. I can begin to make out a covering of fine hairs. Whatever surface they are growing from looks papular and spongy. But I'm drawn to the dark inverted triangle at the top, which looks as pubic as anything in a Palaeolithic fertility figure. Against my will I'm already sexualising the flower, caught up in the *Orchidaceae*'s famously titillating aura. At 100x the blooms are transformed into spangled landscapes. The individual hairs are clearly visible now, and are tipped with iridescence, as if they carried tiny globules of dew. The whole surface of the plant has the glister of organza. And when I shift to the dark patch, which now fills my whole field of view, I spot – and can scarcely believe what I see – two glowing crescents on either edge. They're made up of individual spots of blue, like tiny LED lights. They are the eyes of a malevolent insect from a computer game. I wonder if the light from the microscope lamp is producing these effects, but they are still there when I turn it off. A few minutes later, with the lamp back on, I notice an extraordinary aroma rising from the blooms – musky, sweaty, meaty. Is this the allomone – the chemical which mimics the female wasp's sexual pheromone – made perceptible to the human nose by the heat of the lamp? It's no great surprise when, scanning up to the flower's centre of operations, I spot a minute fly (not the pollinating wasp species) drowned and entombed in the nectary, lured there by the overpowering scents of sex and sweetness.

That afternoon with the microscope was, for me, a journey into orchid inner space. But I doubt that what I saw – the LED eyes excepted – would have been news to any orchid specialist, and it confirmed the modern version of *Ophrys* species' pollination. The male pollinator is attracted initially, and at potentially long distances, by the pheromone-like chemicals emitted by the orchid flower. At close range, by now intoxicated by its own hormones, it mistakes the insectoid flower for a female. Once it has mounted and clasped the bloom, the feel of the artificial fur will be a comforting confirmation, and the wasp's head, pressing into it, will with luck pick up the pollinia.

But wariness about mistaking metaphorical resemblances for real mimicry is still sensible. In 2011 what is believed to be the only night-flowering orchid was discovered in the rainforests of Papua New Guinea.

Bulbophyllum nocturnum has at the centre of its flower a more perfectly mimicked wasp than any *Ophrys* species. But that's not its lure. Dangling round the body are a cluster of dingy-grey tentacles which resemble (and presumably smell like) some of the local slime moulds. They attract night-flying midges which feed on the real slime moulds, and are believed to transfer pollen as they forage hopefully amongst its appendages. (I wonder if this species began as a conventional wasp mimic, developed proto-tentacles as a mutation, and found that these were more effective as pollinator lures than the wasp likeness.)

Underneath their exotic surfaces, all orchid flowers are designed to encourage mobile animals to make physical contact, wriggle in, pick up pollen, and wriggle out again without transferring the pollen to the female stigma of the same plant (cross-pollination generally being better for a population than inbreeding). They are the only group of plants to frequently use sexual attractants rather than nectar alone. The Mexican bucket orchid uses a system of pollination fuelled by one-stage-removed sexual rewards. It exudes a perfume that is irresistible to male euglossine bees – not because it leads them to nectar, but because it mimics the pheromones that the male bees use to attract females during courtship. Struggling to collect this perfumed wax, some of the mob of excited males lose their foothold, and fall into the bucket below – a modification of the lip – which is filled with a colourless liquid. There is only one escape for the sodden bee, via a steep and narrow passageway. En route it must wriggle under two pollen masses hanging from the roof of the tunnel, which, on contact, instantly disengage from the orchid and become attached to the bee. It will be hours or days after the hapless bee and his backpack have emerged before he is attracted by another bucket orchid flower – and will then go through precisely the same ordeal, the insect equivalent of Theseus's trials in the Minotaur's maze. Except that this time the stigma – which lie in front of the pollen sacs in the escape tunnel – snatch the pollen backpack, the orchid is fertilised, and the male bee, once dried out and sober, is able to use the aphrodisiac perfume attached to his legs as part of his elaborate display flight. The South American *Catasetum denticulatum* doesn't even bother with nectarous come-ons. When a bee lands on the lip, it fires a small, winged, cupid's dart from overhead, a pollen arrow, which sticks to the bee's back with a quick-drying glue and remains there until the bee visits another flower.

The extraordinary mechanical contrivances which orchids employ to ensure cross-pollination – rocket launchers, pistons, trap doors, levers, triggers – were another source of their appeal to the industrially minded Victorians. The Madagascan Star of Bethlehem orchid has co-evolved an exclusive oil-drill arrangement with a single species of night-flying moth, which has a *twelve-inch-long* tongue, enabling it to reach the nectar at the bottom of the flower's tunnel-like nectaries. While it is sipping, the moth's forehead comes in contact with the pollen-bearing stamens, whose farina will be transferred to the next flower it visits. When Charles Darwin first saw this flower in 1862, its insect partner wasn't known, and in a letter to his friend Joseph Hooker he puzzled over what kind of creature could possibly probe it. Later that year he published his classic *On the Various Contrivances by which British and Foreign Orchids are Fertilised by Insects, and on the Good Effects of Intercrossing*, and predicted that for the orchid to be polli-nated, there must be some Madagascan insect, probably a moth, whose tongue was a foot long. A few years later Alfred Russel Wallace found an African hawkmoth with an eight-inch tongue, *Xanthopan morganii*, and reckoned that an even longer-tongued relative was entirely possible. Darwin was dead by the time it was found in 1903 but in a nice act of homage, the team who made the discovery named the new super-sipper *X. morganii praedicta* – the predicted one.

What time span of co-evolution can this extraordinary partnership have involved? Were there intermediary stages, short-tongued moths, stumpy nectaries? The benefits to the moth of having a monopoly on this particular food source are obvious (and its telescopic tongue can presumably probe other flowers as well). But what possible advantages can there be for the orchid in making its pollination so exclusive? Why not open house, with multi-species opportunities? The evolutionary answer, of course, is that it simply happened. A moth with a slightly longer than usual tongue found what would otherwise have been an heirless Star of Bethlehem orchid with a slightly deeper than usual nectary. Over millions of years this process was progressively repeated, eventually rewarding moth and orchid with membership of the exclusive Twelve Inch Deep Club.

Darwin was obsessed by the possible dangers of self-pollination. There were personal reasons for this, as he had married his first cousin and fretted throughout his life about the effects this might have on his children. Nonetheless he puzzled about the immense difficulties orchids presented to their cross-pollinators, and admitted that there was much he didn't understand. When it came to the fly orchid he wrote openly that 'what induces insects to visit these flowers I can at present only conjecture'. He could cope with the sexual side of the answer, but maybe not with the idea of such deceit in God's creation. Worse, the laboriously evolved obstacle courses often seemed to work too well, and his careful experiments and observations showed that successful collection of pollinia was far from universal. 'Something seems to be out of joint in the machinery of its life,' he concluded – a rare criticism of the efficiency of evolution. (It is touching to see the great man confess confusion like this, but still keep his essential Darwin-ness, like a duck in winter plumage.) On the occasions orchids are successfully pollinated they produce immense quantities of seed, presumably by way of compensation, though Darwin did not excuse them on this count. He thought it 'a sign of lowness of organisation … a poverty of contrivance'. The bee orchid, inconveniently, seemed to be far more successful by playing the risky game of self-pollination.

Darwin's orchid book is a reverent and acutely observed account of the mechanisms of plant reproduction, and also an honest admission that his theories about the evolutionary benefits of cross-pollination did not always fit the facts of the plant world. More than forty years later, Maurice Maeterlinck put a confidently vitalist gloss on the same processes. He accepts Darwin's theory of natural selection but believes that evolution is driven by some innate and purposeful life force. Also, perhaps, by the plants themselves, using an unconscious 'vegetable intelligence' (and thus prefiguring twenty-first-century discoveries).

In *The Intelligence of Flowers* (1907), he gives a fanciful but botanically accurate account of the pollination of the star species of my valley, the early marsh orchid. It begins with the usual entry of the nectar-seeking insect (often a fly), which inadvertently picks up pollen sacs en route:

> So there we have the insect capped with two upright horns in the shape of a champagne bottle. Unconscious artisan of difficult work, it next visits a neighbouring flower. If the horns remain stiff, they will simply strike with their pollen packets those other packets whose feet are soaking in the watchful stoup, and nothing will issue from the inter-mingling of pollens. Here the genius, experience and foresight of the orchid stand out. It has calculated to the last second the time required by the insect to suck the nectar and move to the next flower, and it had figured this out to be on average a thirty-second interval. We have seen that the packets of pollen are borne on two short stems inserted into the sticky pellets; now, at the point of insertion, we find, beneath each stem, a small membranous disc whose sole function is, after thirty seconds, to contract and to fold each of these stems, so that they describe a ninety-degree arc. It is the result of a fresh calculation, on this occasion not in time, but in space. The two horns of pollen that cap the nuptial messenger are now horizontal and pointing in front of the head, so that, when it enters the next flower, they will strike precisely against the two welded stigmas beneath the overhanging stoup.

An image of Marcel Marceau presses irresistibly into my mind when I read this account of an elaborate act of botanical burlesque. It's intriguing how anthropocentric images and analogies work in Maeter-linck's description. He begins with a clear statement that the orchid is an 'unconscious artisan', but then talks of its 'genius, experience and foresight'. He is being metaphorical, suggesting that the orchid's evolved reflexes seem to show these conscious human qualities. But 'calculated' is a more contentious word, suggesting an ongoing, intelligent process. If the interval between the arrival of the insect and the contraction of the pollen-stem discs is indeed always thirty seconds, regardless of the speed at which the insect conducts its 'difficult work', then the contraction is simply another evolved reflex, ending up with some tardy insects picking up fewer pollinia than they might. But if the orchid is able to alter this time interval to accommodate insect laggards a much more interesting process is going on, which awaits a modern Darwin, or Maeterlinck, for interpretation.

Orchids' aura of innuendo and sexual complexity continues to ensnare their fanciers. A century after Maeterlinck, Eric Hansen, by now as intoxicated by orchids as the enthusiasts he was documenting, sees candidly erotic similes in a *Paphiopedilum* hybrid. The glossy, 'candy-apple-red' staminode that covers the reproductive organs reminds him of the extended tongue famously used by the Rolling Stones as a logo. (Did he know of Joris-Karl Huysman's glossal similes?) 'This shocking red protrusion nestled in the cleavage of two blushing petals then dropped down to lick the tip of an inverted pouch ...' This might seem like the personal fantasy of someone with a unique taste for orchid porn – until you read Hansen's account of American orchid shows, and of the kind of dialogue which passes between the judges:

'Nice lip,' ventured a student judge nervously.

'I've seen bigger,' said one of the accredited judges.

'Big, black and beautiful, but the bugs been at it,' pointed out a third.

'Good gloss, but fatal flaws,' chimed in another.

'Cuppy, cuppy,' a woman snorted.

No wonder John Ruskin, arbiter of Victorian aesthetic taste, was sufficiently repelled to consider orchids 'prurient apparitions'. He also saw metaphorical tongues, though they were lascivious not diseased. One group to which he gave the personal tag of 'Satyriums' (he may have meant the monkey and military orchids – their Tudor name was 'Satyrion') he saw as invariably dressed 'in livid and unpleasant colours' and with the habit of twisting their stalks and lower petals 'as a foul jester would put out his tongue'.

The lust for orchids began during Ruskin's lifetime, when the arrival of a suitable space – the hothouse, the aristocracy's vegetable harem

– coincided with the flowering of the nineteenth century's love–hate relationship with sensuality. Orchids were the perfect *dramatis personae* for the Victorians' vision of nature as a carnival of fabulous, feral and occasionally dangerous wonders, which could be captured and staged as cultural, even national, property. The more spectacular the growths that were brought back from the tropics – each of them a small distillation of these bounteous wildernesses – the more imperial expansion made sense. Orchids' sensuousness and cryptic carryings-on were titillating to the era's repressed desires, and could be made respectable by domestication and display in the enclosures of the Wardian case and the stove. Contemplating the orchids in one contemporary glasshouse, the romantic novelist Charlotte M. Yonge declared: 'Their forms are beyond everything astonishing ... there are hovering birds and very wondrous shapes, so that travellers declare that the lifetime of any artist would be too short to give pictures of all the kinds that inhabit the valleys of Peru alone ...' She was reminded of 'a picture in a dream. One could imagine it a fairy land, where no care, or grief or weariness could come.' Who wouldn't want a piece of that? The Victorians called the craze 'orchidelirium'.

It is no surprise that, as with strelitzia and the Amazonian water lily, rare orchids were deployed in tributes to royalty. In 1837 James Bateman disproved Yonge's belief that orchidaceous luxuriance was beyond the embrace of artists by publishing a monumental illustrated book entitled *Orchidaceae of Mexico and Guatemala*. Bateman was probably the most fanatical of all Victorian orchid collectors, and by the 1840s his home at Biddulph Grange had the most extensive collection in Britain. He described their flowers as 'the chosen ornaments of royalty' and conceived the idea of commemorating them in a book of extraordinary size and elegance, which he dedicated to Queen Adelaide (dowager of the late William IV). The book is a magnificent and lavish oddity. The ravishing illustrations were painted by two self-effacing London women (a Miss Drake, of Turnham Green, and Mrs Withers, of Lissom Grove) over a period of about five years. They are possibly the best botanical studies of the Victorian era and capture the lustre and sculptural intricacy of the real plants. But the text runs like a mischievous eddy alongside, nudging the mainstream text with orchid jokes, improbable traveller's tales, details of native folk dress and cockroaches' food preferences in the glasshouse. The accompanying comic vignettes, by J. Landells, add to the impression that Bateman is gently satirising the Victorian cult of nature at the same time as celebrating it. One of the pen-and-ink cartoons shows an entire carnival of zoomorphised flowers: *Cypripedium insigne* flying out into the night as a witch, a pair of *Masdevallia* dancing a minuet on their leg-like streamers, two *Cycnoches* sailing about on their backs as if they were swans. Bateman joins in the visual punning in his text: '*Cycnoches loddigesii*, perhaps, bears, on the whole, the closest resemblance to the feathered prototype …' But *C. ventricosum* is closest to 'the swelling bosom' of a swan and, if the two species were united, 'we should have a vegetable swan as perfect in all its parts as are the flies and bees with which the orchises of English meadows present us'.

The tone of orchid fancying was not always as entertaining as this. There was snobbery and greed and an often cynical myopia about the fortunes of the plants themselves. Orchidelirium accurately reflected the social values of the people rich enough to indulge in it. The collector Frederick Boyle smugly decided that orchids were 'expressly designed to comfort the elect of human beings in this age', and it was the filling of the elect's customised glasshouses that fuelled the nineteenth-century plunder of a whole botanical family.

Before the 1830s there were comparatively few cultivated orchid species in Britain which had their origins in the tropics. The first to be successfully grown was *Bletia purpurea* from the Bahamas, which was coaxed into flowering in 1731 in the glasshouse of Sir Charles Wager in Fulham. It died soon after. By 1789 there were fifteen species flowering at Kew. They included the huge and dramatic flowers of the *Cattleya* genus, named after the driven collector of exotics Sir William Cattley of Barnet. In the 1830s his field collector found what would soon be called *C. skinneri* deep in the forests of Costa Rica, and brought back a specimen seven feet in diameter, six feet in height, and bearing more than 1,500 flowers, having purchased it, with considerable difficulty, from a local tribe for whom it was a sacred icon. It survived the journey across the Atlantic and, as the largest orchid ever discovered, enthralled the Victorian public when it was exhibited. But the cattleya with the most romantic history is the luscious, frilly, lavender and crimson *C. labiata vera*, which first flowered in Cattley's hothouse in 1818. One of his collectors, William Swainson, had sent him a cargo of plants from the Organ Mountains in Brazil. Out of curiosity, Cattley had planted out the packing material from this collection (botanists often indulge in this obverse of the lucky dip), and from it sprang this prodigious bloom. For a while the plant was the pride of exotic collections, but the specimens gradually died out until only one was left, which was destroyed in a fire. By this time no one seemed sure where the original had been collected. The story of its subsequent rediscovery is possibly apocryphal, but not much out of kilter with the pervasive phantasmagoria of the orchid world. Some seventy years after its disappearance, it re-emerged in Paris, as the improbable but unmissable corsage of a woman at an embassy ball. One of the guests from the British Legation was an orchid enthusiast, and was sure he recognised it. An expert confirmed his guess, and a trail was followed back to the exact site near Pernambuco in Brazil where the plant had been gathered.

By the middle of the nineteenth century orchidelirium was in full swing. The owners of hothouses sent their agents and gardeners across the tropics, to Guatemala, British Guiana, Mexico, Borneo and the Philippines. Extraordinary prices – sometimes up to 1,000 guineas a root – were paid for the choicer plants. The most rapacious collector, the Czech Benedikt Roezl, despatched orchids by the ton, and his cargoes often contained in excess of a million plants. The damage to the orchids' native habitats was devastating. Forests were ransacked. Collectors routinely cut down thousands of mature trees to collect the epiphytic orchids living in their canopies, and often deliberately elimi-nated a species from an area to thwart other collectors and maintain its rarity value. Forged maps were distributed to send rivals on wild goose chases. A photograph from Albert Millican's *The Travels and Adven-tures of an Orchid Hunter* (1891), entitled 'Native Dinner-Time', shows

Snapshot of an orchid collection expedition in Colombia, taken from Albert Millican's account of his travels in the northern Andes (1891) and captioned 'Native Dinner-Time'. The orchid rhizomes are piled centre left.

a party of a dozen native collectors, grouped in a clear-fell round a fire and surrounded by waist-high piles of epiphytic orchids. The Director of the Zurich Botanical Garden said: 'This is no longer collecting. It is wanton robbery and I wonder that public opinion is not stronger against it.' When a specimen of the desirable *Dendrobium schneideri* was offered for sale at auction in London, it had – to fulfil a promise made to the tribe who'd originally owned it – to be sold inside the human skull in which it grew.

Many of the orchids gathered so recklessly died, either on the voyage home or in the unsuitable hothouses to which they were consigned. The Western belief in the ubiquitous virtues of 'fertile' soil, and an ignorance of the complexities of tropical habitats, meant orchids rarely got the growing conditions they needed. In the wild most tropical orchids have evolved to grow without much of what we would regard as humus, living up trees on bark trickle, or in the inch or so of sand that is the ground base of the rainforest. All dead organic matter is rapidly recycled into the seething mass of plant and animal life *above* the ground. In the early days of collecting orchids were often plunged indiscriminately into hot compost, the European gardener's idea of a forest floor simulacrum. No distinction was made between orchids which grew in the ground and epiphytes attached to trees and rocks, and it was not until the 1820s that orchids began to be bought back on the woody substrates they grew on. Before he became involved with Nathaniel Ward's miniature glasshouses, Conrad Loddiges, George's brother, had repeated failures ferrying orchids home by sea. He might have grasped the secret if he had been more attentive of the success of an *Oncidium* from Uruguay, which had flowered repeatedly while hung up in his cabin without any earth, living on air and the condensed moisture in the cabin. Even today the growing shoots of some species are snipped off during transportation because it is not understood that some species send out flower shoots *below* the roots.

Joseph Dalton Hooker, son of Sir William Jackson Hooker, Director of Kew Gardens from 1841, and Regius Professor of Botany at Glasgow University, collected orchids in the Himalayas in 1850. He left a journal, which shows the extent to which even knowledgeable botanists of good pedigree could lose their senses – and their principles – when confronted by these bewitching plants. It's doubtful he was wilfully rapacious or greedy, just that the Victorian imagination could not comprehend the fact that natural resources were finite, and not the a priori legacy of the colonial powers. In the Khasia hills, Hooker found *Vanda coerulea* – that rarest of things, a *blue*-flowered orchid – growing in profusion in woods of dwarf oak. This was an unusual epiphyte, growing exposed to 'fresh air and the winds of heaven ... winter's cold, summer's heat, and autumn's drought [and] waving its panicles of azure flowers in the wind' – conditions completely at odds with the standard methods of cultivation in Britain, which penned all orchid species in hot, airless enclosures. Nonetheless, Hooker picked '360 panicles, each composed of from six to twenty-one broad pale-blue tasselated flowers, three and a half to four inches across'. They were destined for 'botanical purposes' in the inhospitable clime of Britain, and made 'three piles on the floor of the verandah, each a yard high; – what would we not have given to have been able to transport a single panicle to a Chiswick fete!' Hooker adds a footnote which seems to go some way beyond the cause of 'botanical purposes': 'we have collected seven men's loads of this superb plant for the Royal Gardens at Kew, but owing to unavoidable accidents and difficulties, few specimens reached England alive. A gentleman, who sent his gardener with us to show the locality, was more successful: he sent one man's load to England on commission, and though it arrived in a very poor state, it sold for £300 ... Had all arrived alive, they would have cleared £1000. An active collector, with the facilities I possessed, might easily clear from £2000 to £3000 in one season, by the sale of Khasia orchids.'

In the summer of 2013, wooed by the unaccustomed sunshine, I held an orchid show of my own. I was curious about the visual cues that were regarded as the most attractive by commercial breeders; which particular qualities, if any, they selected for. So I bought a few pots from the local garden emporium, at £15 a plant: some *Phalaenopsis* hybrids, from an Australian genus advertised as 'Moth Orchids'; and a 'Spider Orchid' called 'Cambria', which I eventually traced to a man-made group of intergeneric hybrids, originally spliced from two unrelated South American families in the early twentieth century. I put them on my desk, and gazed intermittently at them for weeks.

As whole plants they were ungainly, having, in lieu of a rainforest tree, to be propped up by bamboo canes. They had not an iota of scent (at least not to my nose). The Cambria blooms were undeniably colourful – mottled orange-red petals, speckled peach-skin lips with a golden spot at their top. But they had a kind of stiffness, a too-perfect symmetry, as if they'd been machine cut from dyed fabric, not grown from plant tissue. Put bluebells, or chrysanthemums, or even a flower with sculptural echoes of the orchids like a sweet pea, in an indoor pot and they're animated even when they're motionless. Your eyes track the slight irregularities in the petals, follow fluencies and uncertainties of colour and the ebb and flow of tissue tone as the temperature or humidity changes. Not so with these factory-farmed orchids. They gazed blankly back at me, as frozen as passport photos. I tried setting the pots outside for a week amid marjoram and lavender bushes sagging with bees. Most insects seemed quite indifferent to them. A few small flies settled on one of the Cambria petals, but I think they were just sunbathing on the warm crimson loungers. Pollinating insects – wasps, bumble bees, butterflies – flew by, but seemed to bounce away as they passed close to the flowers, like magnets of the same polarity. Perhaps they did not even recognise them as flowers. Or perhaps the flowers did have a cryptic scent that repels insects of no use to them. Based on this tiny but I don't believe unrepresentative sample, it would be hard not to suspect that the grail of orchid merchants is a bloom which transcends all the unruly quirks and caprices of a living plant and achieves the dazzling rigor mortis of coloured plastic.

By contrast our local wild orchids are models of engagement and high mood. They're fiercely loyal to season and place – but only when it suits them and the creatures they share a home with. Some years they scarcely put in an appearance. In others they're munched up by ponies or shaded out by early grass growth. That spring of 2013, perhaps because of the stresses of bizarre alternations of weather, they'd put on an exceptional show. Rivers of the shade-shifting purples of southern marsh orchids flowed along dried-out dykes on the commons. Bee orchids appeared en masse on a roundabout on the A11. Marsh helleborines, the cutest of British natives, with flowers as frilly as barn-dance frocks, hoedowned in throngs in the sedge fens.

One day in late June I went to a favourite and orchid-rich haunt in the Little Ouse floodplain. Market Weston Fen displays itself like a spontaneous wild version of the 'garden rooms' template. You walk over a pasture towards a narrow gap in a row of willows and gradually, as you approach the reed-fringed bridge, the fen opens out in front of you. The right-hand section is usually the richest, with half a dozen orchid species scattered amongst a mosaic of boggy pools, ringed with sphagnum moss and red rattle and the mildly insectivorous butterwort. But this time what struck me was some curious tuftings in the left-hand sedge bed. They looked like scoops of ice cream perched on top of the fen – *vanilla* ice cream, to be precise. Close-to they were clearly marsh orchids of some kind, maybe twenty-five of them, magnificently lush and tall with flowers whose suffused tones were hard to pin down – cream touched with citrus, lemon-meringue pie, cauliflower florets perhaps … I've no idea why I was gripped by such a rush of edible similes. It wasn't synaesthesia. I wasn't seeing the flowers as tastes. I think maybe they had just touched some long marinated hunger. I'd not seen anything like them before, but they stirred a memory that made my skin prickle, and my mind went back ten years to when I'd first read of that indigenous and apparently extinct 'straw-coloured' orchid – early marsh, subspecies *ochroleuca*. Given what I now knew about the attitudes of avaricious Victorian orchid hunters, my subsequent behaviour was bizarre. I wandered back and forth, glancing down at the plants, keeping my distance, rather like someone who had spotted a ten-pound note on the pavement and wasn't sure whether to pounce. Partly this was due to the lingering (but nonsensical) fancy that getting too close to the anatomical detail of an organism reduces it, disassembles it as a whole living being.

But inwardly I was hopping with delight at the possibility of having rediscovered a returned prodigal, and the niceties of making the flower's acquaintance had been shouldered aside by a sense of trophy hunting that would have done credit to Hooker. I'd commodified the orchid, and knew it. So I got down on my knees and paid it proper respect. I noted every characteristic I guessed might be significant, and began to love it. Close-to, the individual flowers in the spike looked as if they were locked together in some botanical Rubik's cube. I noticed especially the delicious detailing of each flower's lower lip, its skirt, with a raised ridge down the centre and distinct notches in the outer lobes. Perhaps these provided guidelines for pollinators towards the inner chambers of the orchid, which I could just see down the canal at the base of the petals. I raced home and scoured the guides. The plants were indisputably *ochroleuca*. In a state of some excitement I contacted Martin Sanford, author of the magisterial *A Flora of Suffolk* (2010). He confirmed my identification, but gently deflated my fantasy that I'd rediscovered this errant subspecies for Britain. It had gone and come over the years, reflecting the drastic changes in the water table that East Anglia's wet places have suffered at the hands of intensive arable farming. Habitat changes like this are the major threat to orchids worldwide now, far surpassing the comparatively minor impact of modern collecting. *Ochroleuca* had in fact reappeared in the Waveney Valley at Redgrave Fen in 1980. A single flower had been seen at Market Weston Fen in mid June 1995 – same site, same time as my own 'discovery'. Now, as the water levels in more of these special places are under the control of conservation bodies, *ochroleuca* is reappearing, from dormant seeds which have been dreaming under the desiccated peat for decades. In 2012 there were sixty blooms there, just not in flower at the time I'd visited the fen.

Pride in my own apparent cleverness rapidly subsided, to be replaced by pride for the orchid. I'd seen, by pure luck, probably 95 per cent of the entire British population of a plant now on the Critically Endangered List. It is so distinct from and seemingly reluctant to hybridise with its parent species that I imagine it will have full species status soon – if anyone is still bothered about such niceties. It isn't as conventionally beautiful as some of its colourful neighbours, but impressions of beauty

306

spring from many roots in the orchid world: rarity, exotic history, bejew-elled exteriors, fond associations. I find my vanilla ice cream orchid beautiful for another quality, which it shares with my print of Sander's slipper. It has character. It's a survivor, an organism that joins up the dots in the tidal fortunes of East Anglia's wetlands, and the dots in my life, too, from when I first arrived in its ancestral habitat, feeling a mite endangered myself.

THE REAL LANGUAGE OF PLANTS

IN 2014 THE JOURNAL *Current Biology* published an account of some newly discovered properties of the Patagonian vine *Boquila trifoliolata*. They are, to say the least, unusual and appear to violate most existing ideas about plant adaptation and communication. *Boquila* is a thin-stalked woody climber which can spiral from ground level to high canopy, and is endemic to the temperate forests of South America. Its basic foliage pattern is composed of groups of three roughly spear-shaped leaves. What is not supposed to be possible is that these leaves are able to mimic the colour, shape, size and orientation of those of its host trees.

Habitat mimicry isn't unknown in the plant world. *Lithops* species make a good job of disguising themselves as stones in deserts where every item of moisture-bearing food is hunted out voraciously. *Boquila*'s mimicry is unique in that it isn't confined to copying just its immediate environment, or the leaves of a single host species. Its leaves stay within the green-blue spectrum and keep their formation, but as the vine winds through the tree community over weeks and months, the leaves morph to resemble those of each new supporting species, even ones it may never have encountered before. In the space of a few yards the leaves of a single vine can be as smooth as an ivy's, more rounded like box, then bluish and deeply veined, then yellow-green, serrated, oval ended … The Chilean researchers who discovered this mysterious legerdemain, Ernesto Gianoli and Fernando Carrasco-Urra, made a series of photographs of entwined trees, and had to insert arrows to point out which leaves belong to the vine and which to the host trees, so difficult are they to tell apart. They suggest that the purpose of what is essentially an indefinitely flexible and bespoke camouflage technique may be to lower predation by insects, the vine's leaves becoming invisible against the hosts'. But they have no idea how the vine does its trick, except that, in being able to cope with unfamiliar situations, it is demonstrating the first principle of intelligence.

They float a few theories, that the vine's leaves are sensitive to volatile chemicals emitted by the host, that they are somehow provoked into rapid mutation – 'gene jumping' – but neither seems plausible. Maybe there are photo-receptors in the undersides of the leaves that trigger the changes. Plants are rich in cells which are sensitive to light, and they may have been modified in this species to respond to colours and shapes nearby. The caterpillars of some moth species have photo-receptive cells on their stomachs which kindle changes in body colour to mimic the surfaces they're crawling over. You need to find analogies in the animal world to get any kind of grasp on what is happening. Giant squid, for example, seem able to change their colour at will. But the perils of analogy are shown by *Boquila*'s unofficial popular name of 'chameleon vine'. It's a neat tag, but misleading. Real chameleons don't, as is often assumed, change colour to match their surroundings. Their layers of skin are responsive to temperature and to their mood during courtship and displays of aggression, and are modulated by hormone-like chemicals. When they do appear to mimic their habitats it is pure coincidence. (*Boquila* has also been tagged the 'STEALTH vine', after the American spy plane, hinting at the kind of organisation which will be most interested once its transformative mechanism has been unravelled.)

Meanwhile the vine is redolent with biological mystery and possibility. The gap between *Boquila*'s leaves and its hosts' is a literal space, where unknown transactions of light and volatile chemicals and vaporous genes may be in progress. It's also a metaphorical space, containing our existential distance from the plant world and its communication systems. It is hard for us to grasp that there are, for instance, 'scents' that we can't smell, but which plants, noseless and brainless, can. What follows is an account of the growing understanding of plant senses, and what this might say about their status as intelligent organisms.

27

The Butterfly Effect: The Moonflower

IN 1972 THE ENGLISH ARTIST Margaret Mee had been living in Brazil for the greater part of fifteen years. For a scientifically untrained woman from another continent she had made a powerful impact on Brazilian culture. Her exhibitions of paintings of plants from deep Amazonia had caused a sensation. She had been awarded a Guggenheim Fellowship, and had two of her newly discovered species named in her honour. By the end of the 1960s she was one of a small group of scientists and radicals speaking out against the attrition of the forest by mineral companies and loggers and government-sponsored road schemes. Not long after recovering from a long and debilitating bout of infectious hepatitis in 1968 Margaret made a private commitment. She had been an activist in her youth and resolved that in future her explorations and paintings would have a political as well as a botanical purpose: they would become a statement about Amazonia's irreplaceable vegetal energy.

Early in 1972, exploring the river Negro, she painted for the first time a plant that was to become a grail for her, and a symbol of what was happening to the forest. She'd glimpsed the vanishingly rare and legend bound moonflower, *Selenicereus wittii*, five years before, but it had been a poor specimen, budless and bloomless. The plant she found in 1972 also lacked flowers, but was growing in an *igapó*, a patch of flooded forest dizzy with blue orchids, and it gave her new insights into the pattern of tropical plant growth. Even by the standards of Amazonia's flora the plant she saw, and then pictured, was phantasmagoric:

a spill of leaves (in fact flattened stems) sea green and cochineal red, edged with spines and 'pressed as flat as a scarlet transfer against [a] trunk'. *Selenicereus wittii* is a cactus which climbs trees, and Mee figured it like a suit of medieval armour, with a laterally protruding leaf like the beak of a helmet and two more seeming to be laced up round the tree by their interlocking spines. A plant girded against tough times. The lack of flowers was a disappointment, but Margaret knew enough of the scanty information that had been collected about them – that they opened on just one night in the year, were ambrosially scented, and about as different from the palisade of thorny leaves as a kissing gate from a stonewall. It set her on a quest that was to last the rest of her life.

Margaret next saw *Selenicereus* in September 1977, again on the Negro. The *igapó* was dry at this season, and 'after hard walking through dry, tangled wood I found one plant'. She and her boat crew discovered a few more, but most were dying. The surrounding landscape that she'd loved and explored for over twenty years was unrecognisable. The forest on both banks had been burnt off as far as the horizon, and any obstinate relics and regrowth sprayed with chemical defoliants: 'There was not a spot of green to be seen,' she wrote in her journal. 'The trees were sick with bark peeling and rolling off the trunks' – a visual metaphor of unravelling that appears in several of her pictures. What was left was pocked by open-cast bauxite mining and gouged open for new roads to service the palm-oil farmers and gold diggers. But at last she found a single moonflower in fruit, studded with brown orbs held close to the stems, and resonant enough to inspire a full-scale portrait. Her finished picture, dated February 1978, is dramatically different from the cameo of five years before. It has an air of portentousness, with the livid cactus climbing the foreground like an extending periscope. Behind it is a sepulchral forest of ghost trees, almost leafless, the bleached white of the dead trunks brightened only by the vivid 'transfers' of more moon-flower leaves.

By now, she knew something about the moonflower's life cycle, but the scraps of evidence were not much better than tantalising rumour. The fruits form when the winter floodwaters have risen to just below the flowers, suggesting they might be eaten – and the seeds inside dispersed

Margaret Mee in a canoe on the Rio Negro, Brazil, on her 1988 expedition to paint the moonflower.

– by fish. Margaret still did not know what colour the flowers were; she presumed red, like the dominant colour in the leaves. No one knew what its pollinators were, but bats or hawkmoths seemed likely candidates. Margaret's hope to be the first person to paint this prodigal in the wild was now motivated by more than rarity hunting. The moonflower, in its elusiveness and mysterious entanglements, was emerging for her as a symbol of the complex fragility of the Amazonian ecosystem, whose destruction she was witnessing first hand.

Margaret Mee was born in the Chilterns in southern England in 1909, and moved permanently to Brazil in 1952. Her journey from benign European hill country to torrid wetland, a transformative landscape change, may go some way to explaining the piercing, almost extraterrestrial originality of her Amazonian paintings. But she took other legacies with her on the journey to São Paulo, which added edges of steeliness and ecological insight to her pictures. In the 1930s, busking in London as a freelance painter after dropping out of Watford School of Art, she became a political activist, joining anti-fascist rallies (in Germany as well as England) and addressing the 1937 Congress of the TUC on the raising of the school-leaving age. During the war she worked in the drawing office at the de Havilland aircraft factory, north of London, and learned to live with the constant threat of flying bombs – an education in doggedness that would serve her well in the troubled environments of modern Amazonia. Most crucially for her painting, she was accepted as a full-time student at Camberwell School of Art after the war, aged thirty-seven, and began to study under Victor Pasmore, co-founder of the realist Euston Road Group. She recalls his insistent maxim: 'Look at the shapes – fit the shapes between the spaces ...' Perhaps neither of them realised then that this was a fundamental rule of ecology as well as composition: 'shapes' (and organisms) have to 'fit' and adapt to the 'spaces' or niches they inhabit. Margaret's large-scale Amazon painting would become full of cavernous spaces, pregnant with opportunities for life.

Her move to Brazil was almost accidental, initiated by a visit in 1952 to her sister, Catherine, who lived in São Paulo. She fell in love with the country and shortly afterwards her husband, Greville, sold their house in Blackheath and joined her. Margaret began teaching art in the British School at São Paulo, and Greville, a commercial artist with airbrushing skills, found plenty of work in the city's expanding business network. To escape the crowds and heat they frequently hiked in nearby forests and open spaces on the edge of the city. It was on a suburban tramp by an old tramway line that Margaret saw a castor oil plant, with, to her eyes, very curious leaves and fruits: 'It had such wonderful shapes,' she wrote in her diary, 'I sketched it immediately.' From then on, in Greville's words, 'Margaret put aside all other ideas and began sketching and painting flowers.'

It is just possible that during these learning years Margaret had seen the painting of a moonflower in Robert Thornton's famously grandiose book *Temple of Flora* (1804–11). This immense and extravagant folly, which bankrupted Thornton, included a picture of a related species *Selenicereus grandiflorus*, which had been introduced to Britain from Jamaica about 1700. The portrait, with its mixture of Gothic, Romantic and oriental flourishes, catches the feverish mood of early nineteenth-century plant worship, and shows just how much plant painting would develop over the next two centuries. It is the work of two artists. The cactus flower itself, looking as if it has been dipped in honey and dominating the foreground, is by the landscape painter Philip Reinagle. Thornton delegated the background to another small-time artist, Abraham Pether, and it includes an incongruous English woodland, an ivy-clad tower with a clock face showing midnight ('the hour at night when this flower is at full expanse' in Thornton's directorial notes) and a rising full moon behind the trees in the top right-hand corner. The moon, at least, was to make a second appearance in the art of *Selenicereus*.

In May 1988 Margaret returned to Amazonia after a spell in England recuperating from a hip replacement. She was close to her seventy-ninth birthday, but set off by boat from Manaus straight away. Outside the city the forest alongside the river Negro had now almost vanished, devastated by charcoal burners whose plastic-roofed huts were sprouting on land where trees had once grown. It was six hours before they began to see mature forest again on the banks. Margaret had two assistants with her, Sue Loram and Sally Westminster, plus film-maker Tony Morrison and his assistant. They were heading for the *igapó* in which Margaret had last seen a *Selenicereus*, its blooms finished, six years previously. This time the expedition was single-mindedly directed towards recording its flowering.

Margaret's photo album hints at the excitement she was feeling, camped down in familiar territory on what would be her final trip to Amazonia. She is thin but vivacious, dressed in a sensible bush shirt with her waist-length hair done up in a bun. In pictures aboard her boat she has a mannerism which seems oddly characteristic. When she is touching or pointing to a plant, her other hand is always touching or holding on to her favourite straw hat, as if she is using it as an anchor, or lucky charm. Throughout her thirty years in Amazonia she retained an essential and sometimes old-fashioned Englishness. Her journals and photos suggest she got on easily and naturally with the local Indians, especially their children (she had none of her own). They bring her plants, sit in on her painting sessions, try to cut off bits of her thick mane of auburn hair. Margaret, for her part, mucks in with the housework and passes on her European perspective on plants. But unlike many other modern European explorers she doesn't attempt to enter or translate her hosts' complex cosmology, and their view of the forest as a spirit realm. This never seems like a hangover of imperial hauteur or narrow-mindedness, more an honest attempt to keep some clarity of vision inside the cultural language she understood.

Her physical trials in the forest are harrowing. She contracts hepatitis and malaria, and is repeatedly disabled by insect bites (insects, as many painters in the tropics have discovered, also love to eat the paint straight from canvases). She is robbed, swindled, threatened with assault by drunken migrant workers. In the face of this she adopts the unflappable obstinacy of an Englishwoman who has lived through the war. She dispels male predators with mockery, backed up by a small revolver. She sabotages the wildlife shooting sprees of visitors from the towns by literally rocking their boats. She wears pyjamas in her hammock. Holding on to her English hat seems to me to have helped ground her paintings. Although she had an extraordinary, empathetic awareness of the living complexity of the forest, she reveals this through the interplay of space and light, universals of visual language for all species, and never by a descent into mystical imagery.

For the next couple of days Margaret and her team explore the creeks around their base. They find a few moonflowers, but as usual their blooms are over, or the buds blown off the tree. At last they find a spray with two buds *au point*. 'Surely,' Margaret writes, 'it [has] to be the night.' It is an evening of thin cloud, and the low sun is shimmering on the river, turning the forest an aqueous yellow-green. Within an hour the sky is black and only the brightest stars are visible. Margaret and her companions take it in turns to keep watch on the buds, using brief flashes from a torch so as 'not to disturb the opening'. Once, in the confusion of the dark, a torch slips overboard, and they watch its beam slowly dimming as it falls through 'the deep, tea-coloured water'. For two hours the bud doesn't change, just marks time. Then, quite rapidly, the first petal begins to unfurl, and then others, until the whole flower has sprung to animated life. Margaret moves her sketching gear to the roof of the boat, and Sue stands beside her with a pair of battery lights. Tony's camera floodlights are much brighter and Margaret, with no idea of what is supposed to happen next, begins to worry that they are slowing the flower's opening. She asks for them to be turned down to the lowest practicable glimmer, and continues to sketch by light of the full moon which has providentially begun to rise over the dark rim of the forest. Within an hour the pure white bloom, frilled and starry and nearly twelve inches from stalk to petal tip is fully open. The petals are like thin swan's feathers. Everyone is transfixed by its spectral beauty and 'extraordinary, sweet perfume'. But Margaret is still anxious, this time about the absence of pollinators. Her early worries that the lights

317

might interfere with the flowering now grow into a concern that the presence of several busy humans may have similarly disturbed the insects, and that they have 'upset a delicate balance between plant and animal which had taken millions of years to evolve'. She has an image – an epiphany, almost – of such intrusions 'multiplied across all species in the length and breadth of Amazonia', and finds it hard to contemplate. In her mind, it is a real-world corollary of that ecological metaphor about interconnectedness – the flap of a butterfly's wings in Brazil setting off a storm in Europe. Here, the European intrusion might be stopping the insects' pollination flights. She finishes the sketches, and the absence of pollinators becomes one of their potent spaces. Later, in Rio de Janeiro, she turns them into one of the great masterpieces of botanical art, a haunting portrait of the dusky intricacy of the rainforest, an interior knitted from dark spaces and labyrinthine verticals, and lit as much by the four flowers in the foreground as the haze of moonlight behind the trees. Two arm-like flower stalks reach out from the top of the cactus, seeming to support the moon, a vast luminescent disc located in the top right of the painting, just as it is in *The Temple of Flora*'s Gothic illustration.

There were to be some postscripts. A couple of days after the sketches were made it was Margaret's seventy-ninth birthday, and she celebrated the twin achievements with a small party by the side of the Rio Negro. In a snapshot she has exchanged her bush shirt for a chic evening dress with puffed sleeves, and she and Sue sit smiling against the sunset with a bottle of champagne and a cake made of the Amazon fruit cupuaçu. Six months later this woman who had survived fifteen voyages deep into one of the most dangerous environments on earth was killed in a road traffic accident in rural Leicestershire.

Four years after Margaret Mee's vigil on the river Negro, I watched a moonflower open myself – only it was a different, commoner species, and I was at a smart party in Bath, not in an Amazonian *igapó*, adrift

Selenicereus wittii against an Amazonian full moon. Margaret Mee, 1988.

amongst the ghost trees. Our host was an enterprising plantswoman who'd found a 'Night-flowering Cereus' for sale in a garden centre. The description on the label was expansive in praise of its scent, and hinted that the night of flowering was predictable from the swelling of the buds. So she thought a celebration – a floral Opening Night – might be on the cards. Potential revellers were put on twelve-hour telephone alert.

We were duly summoned one evening around cocktail hour, and a few drinks later the buds started to unravel, like inexpressibly slow gasps. Their aroma was extraordinary but disorientating. It wasn't floral at all, but like fruit – luscious, ripe, deliquescing fruit – and resolved itself for me somewhere between pineapple and very heavy melon. In an unintended echo of Margaret Mee's last evening on the Rio Negro we toasted the blooms with an artfully fragrant champagne.

I still have a photograph of the cactus, fully open, taken later that evening. There are six pure white, multi-petalled blooms sprouting from a tumbling, broad-leaved vine, which is set in a terracotta pot, on a newly ironed Provençal cloth, on a wicker table. Twenty-three years on I can identify it as probably a commercial hybrid of *Selenicereus greggii*, a related species from south-western states of North America, where it's known and admired as 'the Queen of the Night'. I doubt we could have imagined that a moonflower drinks party – which felt so chic, so *modern* – had been a commonplace of south-western American life in late Victorian times, when folks used to drink sarsaparilla on their porches as the Queen drowsed out her scent under the desert stars. In fact I doubt if any of us who were present – certainly not me – had given much thought to the plant itself, what sort of creature it was, why it should flower in such an extravagance of scent on just a single night in the year. It was, that evening in an English conservatory, just a piece of especially diverting interior decor.

In this I guess our little company, who would all have classed themselves as plant lovers, were in tune with the twentieth century's floral zeitgeist, and Margaret Mee, with her intense vision of the moonflower as a symbol of connectivity, an organism with a life to lead beyond her needs and human perspective, who was the visionary. It was as if she were an intruder from some alien community in which the plants themselves dictated botanical etiquette. But in the mid 1990s, her notes and drawings and specimens proved to be a vital foundation for the first full scientific paper on *S. wittii*, published by a quartet of German botanists. Almost nothing was known about the plant prior to Mee's explorations, and herbarium specimens were rare and sterile. The German team started work with a collection of clones cultivated in several botanic gardens and one wild specimen from Manaus, where Margaret had painted the plant in the wild. They had also been sent a fresh ripe fruit by her just before her death, plus the crucial information that the plant came into flower at the period when the floodwaters were highest. From these fragments of evidence, and analysis of the fragrance by gas chromatography, they pieced together a biography of what they call, with some understatement, a 'paradoxical life form'.

S. wittii is a true epiphyte, its roots emerging from the flattened leaf-like stems and attaching themselves to trees. They have no connection with the ground. The stems photosynthesise by what is called crassulacean acid metabolism (CAM), not uncommon in epiphytes in hot climates. CAM reverses the usual daily cycle of gas exchange. The pores (stomata) on the leaves or stems, through which gases and water vapour move, open widest at night when temperatures are lowest, to harvest carbon dioxide with the minimum loss of water by transpiration. During the heat of the day the stomata close, lowering water loss but allowing the emission of large quantities of carbon dioxide.

The flowers open for one night only in May, normally at the full moon. The pure white petals reflect the ultraviolet rays in moonlight, making them especially visible to night-flying insects. The heavy scent is also channelled at moths. It has been identified as a mixture chiefly of benzyl alcohol, benzyl benzoate and benzyl salicylate, compounds which, to give a rough idea of where in the scent spectrum the moonflower lies, also occur in jasmine, tuberose, ylang-ylang, hyacinth and balsam. They are typical of what is called the 'white-floral-image', a pattern of sense stimuli to attract night-flying moths. The act of pollination has never been seen, nor the pollinators positively identified, but from the extreme length of the flower and nectar tube (ten inches) they could only be successfully probed by two species of local hawkmoths with tongues of similar length, *Cocytius cruentus* and *Amphimoena walkeri*. These are the south American analogues of the Madagascan hawkmoth Darwin correctly prophesied must exist to pollinate the Madagascan Star of Bethlehem orchids.

The fruits, which ripen after one year, are unusually structured, the seeds being surrounded by large air-filled flotation chambers, so that when they fall into the floodwater, the seeds bob off like corks. Some float away on the current until they are snagged by a host tree trunk; others are eaten by fish, which excrete the seeds, to occasionally end up at the same destinations. They germinate in contact with the trunk, and the cactus can then climb up to six feet or more. But in especially high seasonal floods much of the plant may be underwater, which suggests why the stems have evolved as flat, thick-skinned pads, which wrap themselves tightly round the trunk like a wet suit. When Tony Morrison visited the plant Margaret had painted the following year (1989) he found it entirely submerged by water.

What Margaret had intuited has now been proved. *Selenicereus* is indeed an indicator of forest fragility, a lens on its complexity. It knits together not just the forest's organisms, but its fundamental elements – air, water, seasonality. Moth food becomes fruit becomes raft becomes fish becomes tree climber.

28

The Canopy Cooperative: Air Plants and Bromeliads

ONE OF THE AMAZONIAN SPECIES discovered by Margaret Mee and subsequently named after her was a bromeliad. When she first saw *Neoregelia margaretae* it was high up in the fork of a tree, had no flowers and was swarming with large and ferocious ants. A few months later, revisiting the same plants, she noticed that its heart was tinged with crimson. 'Slowly,' she notes in her journal, 'this centre became larger and more intense in colour. Then, one day, in the heart of this red rosette appeared a colony of small white flowers faintly tinged with pink, emerging from the water which always collects in the cup.' The bromeliad family, more than 3,000 strong, and confined (save for one species) to the New World tropics, are great gatherers of waters. They are epiphytes – that is to say they grow attached to trees and other vegetation by holdfasts but have no roots in the soil. They harvest moisture from the air, condensing mist on spiny protuberances and collecting rain trickle on leaves and in pouches. One common species, Spanish moss, *Tillandsia usneoides*, is a true air plant, picking up water droplets in the atmosphere on minute scales on its surface. In areas of high humidity air plants can grow suspended from telegraph wires.

Margaret's *Neoregelia* is what is called a tank bromeliad. The leaf spirals catch water droplets and guide them down the channels at the central axis of the leaves to a kind of cistern. In some species these tanks can accommodate several litres. They're far from inert bodies of water. They can support entire miniature ecosystems, with resident aquatic plants and animals, even freeloading carnivorous bladderworts which live on the smaller waterborne insects. The tank also absorbs any debris that falls inside, and as this decays it releases nutrients which the bromeliad absorbs through specially adapted hairs at the base of the cupped leaves. They're sociable organisms externally too, forming epiphytic communities with large numbers of other rainforest species – orchids, fungi, mosses, cacti. Together they form clusters whose weight sometimes exceeds that of the tree they're anchored to. The holdfast roots form mats which absorb water, and can build up small caches of humus. Some of this is formed from particles of debris – bark, dead leaves, ant nests – which are washed down from higher in the canopy. But the bulk probably derives from minute particles of soil brought in by wind and rain, including a special supplement blown 5,000 miles across the Atlantic from the deserts of west Africa. Bromeliad leaves also collect minute soil particles, which support colonies of liverworts, lichens, mosses and clusters of their own seedlings, epiphytes growing on an epiphyte.

This is where the feedback loops become really interesting. The supporting trees get paybacks from the relationship. Their trunks are swaddled against dehydration. Quite often they sprout aerial roots whose tips probe down into the epiphytic root mats and suck up moisture. There they build up symbioses with mycorrhizal fungi, which helps them (and the orchids too) extract more nutrients from the accumulations of canopy debris. The entire canopy is an interconnected feeding membrane, endlessly recycling the energy harnessed by the leaves through networks of aerial roots, fungal mycelium and moving water. Insects create other channels for the transport of energy and growth cues. Species which feed on the leaves of one species may pollinate the flowers of another. Others are called in by floral chemicals which aren't nectarous but which mimic the female insect's pheromones, and are then put to work controlling leaf predators. Passionflower vines are common in the New World forests, and one of the groups of insects that feeds on their leaves are caterpillars of heleconids, longwing butterflies. Passionflowers have anciently evolved chemicals in their leaves (glycosides and cyanohydrins) that make them unpalatable. But one insect species, the zebra longwing, has developed resistance, and is not only able to detoxify the chemicals with an enzyme in its saliva, but is able to sequester them for the manufacture of its own poisons, which in turn makes the longwing butterfly toxic to birds. The negotiations don't stop here. In order that its hatchlings have an immediate source of food, zebra longwings lay their eggs directly on passionflower leaves. The vine answers back by producing yellow growths known as stipules on its leaves. They resemble butterfly eggs and are even arranged in similar clusters, to give the impression that the leaf is already occupied. And on the back of the leaf where each stipule has grown there is a small nectar-producing gland, whose aroma attracts ants and wasps. These patrol the passionflower foliage, attacking longwing caterpillars, and providing yet another line of defence ...

The full complexity of canopy life was not understood while it was gazed at from the ground, or hacked down to a human level – as it was in the heyday of orchid hunting. Visitors to tropical forests are generally disappointed by the paucity of dazzling displays of birds and flowers. One modern journalist, Charles C. Mann, travelling in Amazonia and sensing something essential about its cryptic connectivity but not its vitality, remarked that 'to the untutored eye – mine, for instance – the forest seems to stretch out in a monstrous green tangle as flat and incomprehensible as a printed circuit board'. Those who climbed into the tangle, first Indian then European, found something altogether more marvellous. Mark W. Moffett, a Harvard zoologist, performed some of the first systematic climbs in the late 1980s, inspired by a half-remembered science-fiction story from his youth. It featured a world covered with trees so fantastic that the characters walked on the canopy as if it were solid ground.

Moffett's descriptions of his dramatically changed perspective while climbing – the ceiling of the world becoming the floor – are those of someone travelling in a different element. The layers of the forest – broad lower branches grabbing sun specks, towering crowns, vines as trophic spirals of energy between them – remind him of the sea. Through the 'rainforest reef' swim aromatic signalling molecules, minute organisms

Epiphytic bromeliads on a rain forest tree in Ecuador.

like airborne plankton, larger animals flying or leaping from one level to another. This is where the rainforest lives. 'Below, as on the ocean floor, lies a more stagnant, somber world, fed by a rain of dust, feces, branches, leaves and animal corpses ... The trees bridge it all, shifting most organic matter back into their crowns as soon as it becomes available ... to their roots.'

In the mid 1980s, ecologists from the University of Montpellier pioneered a new, low-impact technique for exploring the canopy. Instead of climbing from below, they descended from above, at first using hot-air balloons to settle on the canopy with the minimum of disturbance, then a lightweight raft which could be lowered from a balloon or a more manoeuvrable airship, and give the scientists a platform to work from. It was made from inflatable rubber tubing and nylon netting in the shape of a starfish. The rubber fuselages were broad enough to walk along, and up to five people at a time could scramble about the netting.

One of the great revelations of the new science has been the sheer extent of symbiosis that exists in the rainforest canopy. It's currently estimated that a tenth of all flowering and other vascular plants – some 30,000 species – live as epiphytes in the tropical forests, which means that they live necessarily in complex partnerships with other organisms. Cooperation, or at least negotiation, is the rule. Plants perceive competitors using sense cues – scent, illumination, movement – and coordinate the information to grow away from them. One of the phenomena explorers found most striking is what they poetically dubbed 'crown shyness', the tendency of the growing tips of rainforest trees to be surrounded by a 'personal space' maybe eighteen inches across. In some cases this may be due to wind abrasion between the top twigs but mechanisms to avoid shading by neighbours are also at work.

These mutually beneficial communication networks are ancient and complex. Relationships like that between the moonflower and its specialised pollinator take millions of years to evolve. This is a useful corrective to the declarations of some modern archaeologists that the whole of Amazonia is a 'human artefact'. In the laudable process of highlighting the agricultural achievements of the indigenous Indian populations against claims that the rainforest is a 'pristine' wilderness, they have created another hubristic colonial myth, that human invaders have 'improved' the life systems of countless species, which were in mature social relationships millions of years before the first humans set foot in South America.

The vision of a network of proactive, communicating organisms which has been unfurling through this book – passionflowers with their own pesticides, yew trees morphing their aerial roots into trunks, carnivorous species with the powers of muscled animals, orchids mimicking insect pheromones, arums able to raise their internal temperature – is the newly realised face of the once supposedly passive vegetable world. But what this says about the kind of organisms plants are continues to be contentious, because it raises that ancient conundrum of whether plants can be said to possess intelligence.

29

Plant Intelligence: Mimosa

THE SENSITIVE PLANT, *Mimosa pudica*, is a widespread annual weed of the New World tropics. It is a member of the bean family with delicate, daisy-like purple flowers. But it is the pinnate leaves – deeply and multiply divided, like feathers – which give the plant its common English name. These fold up at night, or whenever they are touched in any way, a fact that was known to colonial explorers, who were circulating specimens amongst European naturalists and gardeners by the early seventeenth century. *Mimosa*'s sensitivities – the closure is a decisive and orderly wave that ripples up the leaf – made it a prime piece of evidence in the argument about the differences between animal and vegetable creation a hundred years before the discovery of the Venus flytrap. And in one way it was more challenging than the tipitiwitchet, whose need to catch insects at least gave a reason for such strikingly unbotanic movement. There is still no satisfactory explanation as to why the sensitive plant shuts its leaves in such a dramatic fashion. But from the mid 1600s there was no shortage of theories as to how it did the trick, which track the diverse mindsets of contemporary 'natural philosophers'. Early speculations were mostly based on mechanical models. In 1661 the microscopist Robert Hooke put the plant through a series of physical trials, analogous to those Charles Darwin undertook with sundew two centuries later. He found that striking or cutting a leaf, or treating it with acid, would cause the lobes above the point of trauma to close, and those below to collapse pair by pair down the leaf. His dissection under the microscope

'Emma Hamilton [Admiral Nelson's mistress] in an attitude towards a mimosa plant, causing it to demonstrate sensibility', George Romney, 1789.

had shown the presence of moisture conduits in the leaf, and he suspected that a blow forced 'liquors' out of the leaf and into the rib, causing the lobes (known as pinnulae) to come together. Philip Miller's bluff summary of the various mechanical and hydraulic theories in *The Gardeners Dictionary* (1731) reads like a plumbing manual. But he prefaces it with a more Romantic meditation: 'at all Hours of the Day, with the least Touch of a Stick, or gentle one with the Hand, the Leaves, just like a Tree a dying, droop, and complicate themselves with all the Quickness imaginable, and presently after erect and recover themselves, returning to their former Position; so that a Person would be induced to think they were really endow'd with the Sense of Feeling'. Some precocious observers thought that electricity – the newly discovered magical 'fluid' – was involved. John Freke, an English surgeon-scientist, assumed that 'the nature of the sensitive plant is to have more of this fluid in it than there is in any other plants or thing'. He thought that the plant was charged with static electricity, and that touching the leaves earthed and discharged it, causing the leaves to hang 'in a languid state'.

The existence of an a priori tendency towards 'irritability' was the most popular speculation, as it was with carnivorousness in plants, and one which emerged from the popular contemporary belief that there was no essential difference between animal and plant tissue. George Bell, in 1776, declared of mimosa and dionaea: '[T]hat these plants *live*, will be granted; but I suspect that they likewise *feel*. I doubt whether we are right in confining the capacity of pleasure and pain to the animal kingdom' – and suggesting, to any who might think his ideas too impractically metaphysical, 'This view of the life of vegetables raises botany to the rank of philosophy. It adds fresh beauty to the parterre, and gives new dignity to the forest.' Erasmus Darwin, who prodded the sensitive plant as eagerly as he did the Venus flytrap, saw both mechanical and 'irritable' processes at work in its reactions. In the verse sections of *The Loves of Plants* he compares the shrinking of the leaves to the falling of the column of mercury in a barometer:

> So sinks or rises with the changeful hour
> The liquid silver in its glassy tower.

So turns the needle to the pole it loves,
With fine librations quivering, as it moves.

But in the attached note he wonders if the reaction may not 'be owing to a numbness or paralysis consequent on too violent irritation, like the fainting of animals from pain or fatigue'.

When the sequence of events which produces the sudden infolding of mimosa's leaves was eventually mapped in full two centuries later, the pitfalls inherent in regarding plants as simplified animals became clear. But the analogists hadn't been on entirely wrong tracks. Electricity was involved, but not as some kind of static discharge. So was the movement of fluids, though not as circulating sap or brute fluid pressure. Sensors which register blows or shaking or any kind of deformation are located in the axes of the leaflets, and when they are activated they provoke a rapid increase in the permeability of the cell walls. Water and electrolytes pour out into the intercellular spaces. This causes a loss of turgor, or rigidity, and the touched leaflets collapse, setting off an electrical signal that is transmitted along the leaf rib, triggering similar collapses downstream. Recovery can only happen in light, which activates the chemical pumps that push electrolyte solutions back into the cells. So, we have a response which is irritable, mechanical, chemical and electrical all at once – the complex make-up of all living processes, and a reasonable answer to how mimosa does it. But *why* it should have evolved its characteristic tic is still unknown. One theory is that its purpose is the dislodging, or frightening, of insects which alight on the leaves. But it seems too gently choreographed to be effective at either. Hurling predators away by more rapid and violent convulsions is entirely within the bounds of physiological possibility for a leaf, but isn't an option that has been taken up in the vegetable world.

Curiosity about *Mimosa pudica*'s abilities went off in an entirely different direction in 2013. Monica Gagliano is an Australian ecologist who has developed a special interest in plant communication and problem solving. Working in the laboratory of the 'plant neurobiologist' Stefano Mancuso in Florence, she set up an experiment to test whether anything resembling habituation happened with mimosa.

Habituation in a biological sense is the recognition and filtering out of irrelevant stimuli, and, in the more popular usage, becoming immune to the effects of repeated stimuli. Gagliano potted fifty-six similar plants, and rigged up a system to drop the pots from a height of six inches every five seconds. Each 'training session' involved sixty drops. All the specimens, to varying degrees, shut their leaves in the proper manner of their species. But some started to reopen after only four or five drops, as if they had judged that the drop-and-jolt stimulus was one they could safely ignore. Were the plants simply suffering from vegetal fatigue, their cell walls overstressed? Apparently not. When the same individuals were subjected to manual shaking their leaves immediately closed. But put through the drop test again, none of them opened. Gagliano repeated the experiment, using the same 'trained' plants, after one week and then one month. Again they ignored the drop stimulus, suggesting they'd 'remembered' what they had learned. Bees, in analogous experiments, forget what they have learned in forty-eight hours.

In the absence of any explanation as to how a plant, with nothing comparable to a brain, could process or store memories, Gagliano's findings have been summarily dismissed by orthodox botanists. It is not so much her experimental data that raises hackles as her conceptual framework and its attendant language. 'It's not learning' is a typical response. 'Animals can exhibit learning, but plants evolve adaptation.' To which Gagliano's reply is 'How can they be adapted to something they have never experienced in their real world?'

The scepticism about mimosa's capacity to learn and remember is an example of a growing stand-off between traditional plant science and the new discoveries about vegetable behaviour (for example *Boquila trifoliolata*'s ability to mimic the foliage it creeps over). Traditional botanists are apt to pour scorn on suggestions that vegetal life can transcend its stereotype of passive immobility. The greatest hostility is reserved for the use of the word 'intelligence' in connection with plants. The vegetable world's florid range of behavioural abilities matches that of many animals, but can't, conventional wisdom insists, be called intelligent without the existence of a central brain to coordinate information, control and modulate behaviour and occasionally override pure reflexes. The concerns have deep roots, in the debates of the eighteenth and nineteenth centuries and the claims that plants had 'feelings' and maybe a primitive kind of consciousness.

And the world of plant science hasn't forgotten a more recent and sensational presentation of the idea of plant intelligence. In 1973 Peter Tompkins and Christopher Bird's *The Secret Life of Plants* became a bestseller, with its anecdotes of how plants felt pain, could distinguish between Beethoven and The Beatles, and ached to communicate wisdom to environmentally negligent humans. The book's most notorious story concerned an ex-CIA polygraph expert, Cleve Backster, who on a whim hooked up a galvanometer to the leaf of one of his office plants. To his amazement he found he was able to make the needle on the galvanometer kick just by imagining the house plant being set on fire. A string of surreal tests followed, supposedly establishing the plant's ability to read human thoughts over long distances. In one, a plant that had witnessed the 'murder' of another of its kind (by stamping) was able to pick out the culprit from an identity parade of six suspects by precipitating a storm of polygraph activity when the murderer was brought before it. Neither this, nor most of the other outlandish experiments, proved reproducible in more stringent circumstances, and the book left an indelible stain of kookiness on the idea of vegetal 'thinking'. For the next thirty years it would be bracketed with other paranormal enthusiasms like UFOs and telekinesis.

Meanwhile evidence of plants' sensory abilities is proliferating. Electrostatic charge recordings have shown that a pulse passes between a flying insect (negatively charged) and a sun-warmed flower (positive), providing yet another landing signal. The volume of moisture that evaporates above liquid nectar is a signal which can be 'read' by insects and indicate whether there is enough left to warrant the expenditure of more insect energy and risk neglect of other flowers needing pollination. The nectar itself may contain caffeine, as addictive to insects as it is to humans and which encourages them to remember a particular plant and return to it.

Chemical communication between plants is proving to be as

sophisticated as that between plant and insect. An expanding tally of species have been shown to 'warn' their neighbours of insect attacks by emitting pheromones, which raise the production of bitter tannins and other predator-deterrent chemicals in nearby plants. Sage brush in North America and willow and oak trees in Europe do this with airborne chemicals (they've been collected in plastic bags and analysed; there are more than a hundred different compounds emitted, including aspirin's volatile cousin, methyl salicylate). So do mopane trees in Africa – though in this case the predators are elephants, which are bright enough to argue back. The browsed mopanes' airborne warnings are blown downwind, so the elephants move *upwind* through the trees, taking just a few leaves from each and hoping to outrun the vaporous messaging. If all of this looks suspiciously like that bête noir of Darwinian thought, altruism, ecologists believe that the warning pheromones are initially emitted as a fast-track signal to the original victim's own outer leaves, and are 'eavesdropped' by neighbouring plants.

But something more indisputably communal is happening underground, in the bush telegraph of root systems. It's long been known that the root systems of most land plants – trees especially – grow in intimate and mutually beneficial entanglement with what are called mycorrhizal fungi. The fungi, which have no chlorophyll, take the sugars they cannot produce themselves from the plant's roots. The plant in its turn absorbs minerals and other nutrients from the fungal network, which is more efficient at extracting them from the ground. What has only recently been established is the fact that, in a forest for example, these networks can link trees of many different species, forming not just a feeding cooperative but a vast signalling network – what one forest ecologist called a 'wood-wide web'. Suzanne Simard, of the University of British Columbia, has injected firs with radioactive carbon isotopes and tracked their progress through the forest with a Geiger counter. Within a few days every tree inside a thirty-yard square had been connected to the network. From the pattern of nutrient traffic it looks as if the more mature trees were using the net to nourish the most shade-compromised seedlings, including those of other species like paper-bark birch. Simard believes that a kind of symbiosis-by-proxy exists between the fir and the deciduous birch with the fungal network acting as an intermediary. The evergreen helps top up the birch's sugar reserves during cold months, and is repaid during the summer. I suspect that the mycorrhizal fungus is also a driver, since its interests are best served if all its tree partners are thriving. (Fungi are in a different kingdom from plants, but their lives are just as extraordinary. In the main reactor at Chernobyl a group of fungi are flourishing by using melanin to convert gamma radiation into living tissue, just as plants use chlorophyll to fix solar energy.)

The last plant communication frontier may be bio-acoustic. The notion of plants reacting to sounds and even emitting them meaningfully takes us perilously close to *The Secret Life of Plants*, and the evangelical Dorothy Retallack playing Bach and Led Zeppelin to her geraniums. (She was trying to prove the morally corrupting influence of heavy rock.) Producing meaningful sound waves would entail a huge investment of energy and specialised cellular apparatus at both the transmitting and receiving ends. It isn't clear that plants have either, nor would have needed to evolve them, given the sophistication of their chemical signalling systems. But sound communication at a short distance might be feasible. Monica Gagliano, who is pushing at many boundaries in the plant intelligence (PI) field, has found that the roots of young corn plants can act acoustically both as transmitters and receivers, emitting frequent clicking sounds and bending towards the source of sounds of particular frequencies. In Florence her colleague Stefano Mancuso has videoed the growth of young bean plants towards a pole, taking a still shot every ten seconds over two days. Contrary to the popular assumption that tendrils weave aimlessly about until they find something to cling to, Mancuso's sequences show a purposeful stretching aimed directly at the pole. They 'know' where the pole is before they reach it. Mancuso believes they may be using echo location, generating similar clicks to corn roots – though how the reflected sound might be picked up and interpreted is difficult to envisage.

The vegetable world has an extraordinary range of sensory powers. But does an individual plant's ability to coordinate the signals it is receiving, from volatile chemicals, gravity, light and maybe even sound, qualify as intelligence, even when it seems to lead to problem solving? Or is it best regarded as chemical computation, made to seem clever by evolution's long programming? As usual, it depends what you mean by intelligence. Traditionalists accuse the New Wave of anthropomorphism, of interpreting complex plant communication as if it were intentional and adaptable, as it is in brain-possessing humans. The philosopher Daniel Dennett, in a neat parry, has labelled such views 'cerebrocentric' and lamented the fact that we find it difficult (and maybe humiliating) to conceive of intelligence as existing in any form other than our own brain-and-neurone variety. 'The idea that there is a bright line,' he says, 'with real comprehension and real minds on the far side of the line, and animals or plants on the other – that's an archaic myth.' Gagliano agrees, and suggests that 'brain and neurones are a sophisticated solution but not a necessary requirement for learning'. Perhaps we should be prouder of that considerable part of our own intelligence which is unconnected with self-awareness, and is the product of long evolutionary processes analogous to those that govern sensible behaviour in plants. Intelligence is simply the ability to solve problems and adapt to changing circumstances, and plants may have their own specific 'thinking' processes. Charles Darwin, certainly no prototype New Ager, thought that the control centre might be located in the growing tip of a plant's root. In the final sentence of *The Power of Movement in Plants* (1880) he says: 'It is hardly an exaggeration to say that the tip of the radicle having the power of directing the movements of the adjoining parts, acts like the brain of one of the lower animals ... receiving impressions from the sense organs and directing the several movements.' And deep in their cellular chemistry plants have been found to have many of the same neurotransmitters and receptors as humans, including glutamate, GABA and cannabinoids.

That plants contain receptors for substances such as cannabis which in humans influence mood isn't really a surprise. These molecules are efficient messengers in living cells, reporting stress and damage and switching on repair mechanisms. Evolution tends not to waste a good invention, and when chemical messengers became necessary for the sophisticated nervous systems of humans, some ancient vegetal patents were called up as prototypes.

And here, ironically, I find myself on a path which I'm not able to follow much further. It's not simply that I'm at the very edge of my scientific competence. Nor that I feel these revelations of deep biochemistry 'disenchant' the plant kingdom in some vague way. For me, discoveries about the complicated and proactive sensory systems of plants dignify them, and take them out of the passive role of 'service providers'. In the future I suspect it will also help move them to a new existential level, as ecological teachers, in what is called, rather piously, bio-inspiration. We may learn from them how to deter insect predators with minute pulses of synthetic pheromones, how to switch on drought resistance in trees using low-frequency sound signals, how to achieve the grail of artificial photosynthesis as an energy source. But this is all in a parallel universe. Just as quantum physics and string theory have no relation to our everyday experience of the material world, so the discoveries of the new botany don't have an impact on how we relate to living plants, beyond validating them as autonomous beings. I rejoice in the fact that we share hormones and sensory skills, but hope that we can respectfully maintain the independence of our kingdoms, not as irrevocably separate orders, but as two linked pathways to the challenges of being alive.

Keats advised against 'hurrying about and collecting honey, bee-like, buzzing here and there impatiently from a knowledge of what is to be arrived at', and that instead we should 'tak[e] hints from every noble insect that favours us with a visit'. The new cabaret of vegetal behaviour seems to me analogous to the hints of bees, to be taken sympathetically but not allowed – 'impatiently' – to reduce our image of plants to aggregates of neurotransmitters and signalling networks. That is how we lost hold of them at the beginning of the twentieth century. What we experience, out in the world – the opportunist weed scent-marking its personal space, the bee-whispering of summer blooms, the veteran tree recycling its energy in the optical illusion of the Fall – are molecular transactions expressed in parameters of existence which we all share and comprehend: occupation of space and seasonal renewal.

Epilogue: The Tree of After Life

THEY WERE THE OTHER POLE of my dawning understanding of plants, the converse of samphire's urgent and transient flitting. Beech trees defined the landscape I grew up in, studding the local commons, hanging on steep chalk hillsides, big trees gracing big front lawns. For me they made a kind of timber frame, but it seemed sinuous and playful and not at all stolid. Beeches move with the weather and, fecklessly rooted, are often felled by it. They're fluid, feline, shape shifting in response to changes in forest ambience. For twenty years I had custody of a wood in which there were a large number of beeches, and they always seemed like sprites, not part of the working proletariat of oak and ash. Thirty-five years after the book inspired by samphire I wrote another, *Beech-combings*. Its star was a giant 400-year-old pollard called 'the Queen' which I had known most of my life, and eulogised like a lovelorn admirer: 'a great inverted bowl through the mist of foliage,' I wrote rhapsodically, 'it still takes my breath away, the mass of it, the hunched shoulders, the low spreading skirt ... its long low branches trail out like the arms of a giant squid'. This magnificent tree survived all the great storms of the 1980s and 90s, and I thought its immense cantilevered branches and low centre of gravity would see it through another few centuries. But in June 2014 I heard from an acquaintance that it had been blown down in a minor gale. He wondered if I wanted to pay my respects.

I'm there a few days later, and unsure for a moment whether I've found the right tree. The space it occupied has changed dramatically, not just because the Queen is now cleaved down the middle and lying like two huge spills on the forest floor, but because of the light, reflected off the gashed wood and last year's crisped leaves and tree surfaces that have not seen the sun for a hundred years. I don't have to look closely to see why, in the end, the Queen tumbled down so easily. Its interior is as crumbly as old cheese, broken down by a wood-eating hoof fungus, and what I'd always thought of as an indomitable matriarch was ready to fall apart with the lightest of summer winds.

What is striking is that the fallen Queen still has total command over the space it occupied while still vertical, and will continue to do so into the future. I don't think this is just because it also has a lifetime's occupation of my memory. The surrounding and still standing trees show the hollows in their crowns where they were shaded out, and which will become the templates for new growth. And a huge bank of data has fallen with the Queen. I find an area high on the trunk where the bark has been ripped off, and underneath the timber is inscribed with the beautiful pattern known as spalting, a lacework of dark lines that mark the frontiers in a long border squabble between another wood-rotting fungus and the tannin defences secreted by the tree. Close to are scores of ancient graffiti, hard to see when the tree was upright. One reads *18. V. 44*. The letters are stretched enough to be Victorian, but this is almost certainly from 1944, when homesick US airmen were stationed nearby, and carved their names and townships on several of the beeches.

The National Trust has put up a notice a little distance from the wreckage with the words 'This famous tree has entered the next stage of its life' – an ecocentric sentiment that would have been unthinkable on public display even ten years previously. But it's a recognition of what is going to happen next. Beyond the Queen's space there are dense patches of birch and young beech marking the sites of earlier pollard collapses, each one with the fallen tree surviving inside it as a dark, shrinking shell, home to fungi and wood-boring insects. This is what the Queen will become in a few decades. On the ground her roots are visible on the surface, as hard and rippled as limestone rocks. They remind me of stromatolites, the compacted clumps of silt and cyanobacteria that were some of the first living communities to form on dry land more than

The Queen Beech, Ashridge, Hertfordshire. Before it fell, the tree was the model for 'the Whomping Willow' in the Harry Potter films.

3 billion years ago. Beneath them the complex mycorrhizal fungi that connected the Queen to the rest of the wood will be biding their time, waiting to hook up with the roots of first new beech seedlings – some of them maybe the Queen's own descendants.

What is the lesson of this recumbent giant? Perhaps just the old cliché 'The Queen is dead. Long live the Queen.' That the original tree is no longer living is beyond doubt. Its tissues are moribund and decaying. But the legacies it has left in this corner of the wood – topographical, architectural, ecological, genetic – are ineradicable, except by some act of gross human intervention. Trees are used to catastrophes, big and small. They have been tacking around them for millions of years, living as they do in complex and mutually supportive communities, not as isolated individuals. At this moment their greatest threats are exotic diseases and predatory insects to which, as yet, they have no intrinsic resistance. This is what is happening with ash trees across Europe and North America. How the ashes themselves will adapt to their new travails is uncertain. But the message of this book has been that plants are never simple victims, passive objects, but vital, autonomous beings, and that listening to and respecting that vitality is the best way we can co-exist with them, and in their difficult times, learn to help them.

Additional References and Sources

Sources referenced in the text are not repeated here.

Introduction: The Vegetable Plot

p. 1 'Just before he died': Edward Lear, *The Cretan Journal*, ed. Rowena Fowler, Dedham: Denise Harvey & Co., 1984.

p. 5 'It's odd that we haven't': Rationales for plant conservation, see: *Global Strategy for Plant Conservation:A Review of the UK's Progress towards 2020*, Salisbury: Plantlife, 2014; Tony Juniper, *What Has Nature Ever Done for Us? How Money Really Does Grow on Trees*, London: Profile Books, 2013; George Orwell, 'Politics and the English Language', 1946, in *Collected Essays, Journalism and Letters,* ed. Sonia Orwell and Ian Angus, Vol. IV, London: Secker & Warburg, 1968.

p. 7 'the Language of Flowers': For a sensitive modern fictional exploration, see Vanessa Diffenbaugh, *The Language of Flowers*, London: Macmillan, 2011.

p. 8 'The great Romantic lover': Coleridge: in *Collected Letters of Samuel Taylor Coleridge*, Vol. II, *1801–1806,* ed. Earl Leslie Griggs, Oxford: Clarendon Press, 1956.

How to See a Plant

p. 10 'The [lotus] emerges': Zhou Dunyi's homily and other lotus lore is quoted in Jennifer Potter, *Seven Flowers and How They Shaped Our World*, London: Atlantic Books, 2013.

p. 10 'Vegetal hygiene becomes': The nano-technology of lotus leaves: see Peter Forbes, 'The Gecko's Foot: Bio-inspiration – Engineered from Nature', *Nature*, no. 7065, 2005, p. 166.

1. Symbols from the Ice: Plants as Food and Forms

p. 13 'The compelling images': For overviews of ice age art see Paul G. Bahn and Jean Vertut, *Journey through the Ice Age*, London: Weidenfeld & Nicolson, 1997. Also the first chapters of N. K. Sandars, *Prehistoric Art in Europe*, 2nd edn, Harmonsworth: Penguin, 1985.

p. 17 'His own interpretation': Tony Hopkins, personal communication. See also his *Pecked and Painted: Rock Art from Long Meg to Giant Wallaroo*, Peterborough: Langford Press, 2010.

p. 18 'Jill Cook, senior curator': Jill Cook, personal communication, and *Ice Age Art: The Arrival of the Modern Mind*, London: British Museum Press, 2013.

p. 20 'The poet Kathleen Jamie': Kathleen Jamie, *Guardian*, 16 February 2013.

p. 20 'When Paul Bahn and Joyce Tyldesley': Joyce A. Tyldesley and Paul G. Bahn, 'Use of Plants in the European Palaeolithic. A Review of the Evidence', *Quaternary Science Reviews*, Vol. 2, 1983, pp. 53–81.

p. 23 'The first botanic gardens': Early botanic gardens: see John Prest, *The Garden of Eden: The Botanic Garden and the Re-creation of Paradise*, New Haven, Conn., London: Yale University Press, 1981.

p. 26 'D. H. Lawrence, continuing': D. H. Lawrence, *Reflections on the Death of a Porcupine and Other Essays*, ed. Michael Herbert, Cambridge: Cambridge University Press, 1988.

p. 27 'John Berger has written': John Berger, 'Painting and Time', *The White Bird: Writings by John Berger*, ed. Lloyd Spencer, London: Chatto & Windus, 1985.

p. 28 'I'd met Tony Evans': Tony Evans's career: *Tony Evans: Taking his Time*, ed. David Gibbs, David Hillman and Caroline Edwards, London: Booth-Clibborn, 1998.

p. 30 'collaborate on a book': Our book, *The Flowering of Britain*, was published London: Hutchinson, 1980, and made into a BBC documentary two years later.

p. 34 'The concept of ancient woodland': Ancient woodland: see Oliver Rackham's work, e.g. *Trees and Woodland in the British Landscape*, London: J. M. Dent, 1976; *Ancient Woodland: Its History, Vegetation and uses in England*, London: Edward Arnold, 1980; *Woodlands*, London: Collins, 2006.

p. 38 'On 24 April 1802': Early records of the daffodil: *A Seventeenth Century Flora of Cumbria: William Nicolson's Catalogue of Plants, 1690*, ed. E. Jean Whitaker, Durham: Surtees Society, 1981; Dorothy Wordsworth's journal from *Home at Grasmere: Extracts from the Journal of Dorothy Wordsworth (Written Between 1800 and 1803) and from the Poems of William Wordsworth*, ed. Colette Clark, Harmondsworth: Penguin, 1960.

p. 38 'The Glow-worm Rock primrose': Molly Maureen Mahood, *The Poet as Botanist*, Cambridge: Cambridge University Press, 2008. For a fuller account of the life-story of 'the primrose of the rocks', see Lucy Newlyn, *William and Dorothy Wordsworth: 'All in Each Other'*, Oxford: Oxford University Press, 2013.

Wooden Manikins: the Cults of Trees

p. 41 'It's on a Sumerian seal': Sumerian seal: Anne Baring and Jules Cashford, *The Myth of the Goddess: The Evolution of an Image*, London: Penguin, 1993.

p. 42 'As for the Tree of Life': 'Trees of Life: see Marina Warner, 'Signs of the Fifth Element', catalogue for *The Tree of Life* touring exhibition, South Bank Centre, 1989.

p. 42 'In the early 1970s': Allen Meredith: see Anand Chetan and Diana Brueton, *The Sacred Yew*, London: Arkana, 1994.

p. 48 'If the Yew be set': Robert Turner, *Botanologia the British Physician*, 1664.

p. 48 'A German medical professor': Dr Kukowka's experiments are quoted in Edred Thorsson, *Futhark: A Handbook of Rune Magic*, York Beach, Me.: Weiser, 1984.

p. 49 'Anglo-Saxon boundary descriptions': Anglo-Saxon boundary trees: see Rackham, *Trees and Woodland*, op. cit.

p. 50 'But the yew which trumps': Fortingall yew modern mythology: see Chetan and Brueton, *Sacred Yew*, op. cit.

p. 54 'What can be more pleasant': John Worlidge, *Systema Agriculturae*, London, 1669.

p. 54 'The Fortingall yew was': Fortingall yew early measurements: see Jacob George Strutt, *Sylva Britannica, or Portraits of Forest Trees*, London: Longman, Rees, Orme, Brown and Green, 1830, Gilbert White, *The Natural History and Antiquities of Selborne*, London, 1789; Patrick Neill, quoted in Chetan and Brueton, *Sacred Yew*, op. cit.

p. 58 'in the centre of the original tree': J. E. Bowman, 'On the Longevity of the Yew', *Magazine of Natural History*, NS 1, 1832, 28–35.

p. 58 'But the weight': John White, 'Estimating the Age of Large and Veteran Trees in Britain', *Information Note,* Forestry Commission, November 1998.

4. The Rorschach Tree: Baobab

p. 65 'At Ifaty on Madagascar': Famous baobabs: see Thomas Pakenham, *The Remarkable Baobab*, London: Weidenfeld & Nicolson, 2004; Michel Adanson, *Histoire naturelle du Sénégal*, Paris, 1757.

p. 69 'The West African writer': Seydou Drame, quoted in Pakenham, *The Remarkable Baobab*, op. cit.

The Big Trees: Sequoias

p. 72 'In the New World': Sequoia history: see James Mason Hutchings, *Scenes of Wonder and Curiosity in California*, 4th edn, San Francisco, 1870; Simon Schama, *Landscape and Memory*, London: HarperCollins, 1995.

p. 73 'It is to the happy accident': Josiah Dwight Whitney Jr, *The Yosemite Book*, New York: Julius Bien, 1868.

p. 73 'Another writer': Charles Fenno Hoffman, quoted in Roderick Frazier Nash, *Wilderness and the American Mind*, New Haven, Conn., London: Yale University Press, 1967.

p. 76 'In 1901 another American': Details of the Roosevelt-Muir meeting in John Muir, *Nature Writings,* ed. William Cronon, New York: Library of America, 1997, Robert Macfarlane, *Landmarks*, London: Penguin, 2015.

6. Methuselahs: Bristlecones and Date Palms

p. 79 'In the 1960s a Michigan': David Milarch's story and a discussion of 'Champion Trees' is in Jim Robbins, *The Man Who Plants Trees*, New York: Spiegel & Grau, 2012.

p. 80 'The longest proved dormancy': Date-seed dormancy: see Thor Hanson, *The Triumph of Seeds*, New York: Basic Books, 2015.

p. 82 'The answer given': Chris Walters quoted in Hanson, *Triumph of Seeds*, op. cit.

7. Provenance and Extinction: Wood's Cyad

p. 88 'an Eden of the remote past': Oliver Sacks, *The Island of the Colour-blind*, London: Picador, 1996.

8. From Workhorse to Green Man: The Oak

p. 90 'Historical ecologist': Oliver Rackham and A.T. Grove, *The Nature of Mediterranean Europe: An Ecological History,* New Haven, Conn.: Yale University Press, 2001.

p. 91 'William Bryant Logan': William Bryant Logan, *Oak: The Frame of Civilization*, New York, London: W. W. Norton, 2005.

p. 92 'as Rackhan remarks': Rackham, *Trees and Woodlands*, op. cit.

p. 92 'The naval historian': John Charnock, *A History of Marine Architecture*, 3 vols, London: R. Faulder, 1800–1802.

p. 96 'If I'd had more': Form and pattern in tree growth, see: Philip Ball, *The Self-made Tapestry:Pattern Formation in Nature,* Oxford: Oxford University Press, 1999; D'Arcy Wentworth Thompson, *On Growth and Form*, Cambridge: Cambridge University Press, 1961.

p. 97 'In 2011, the French physicist': Christophe Eloy, and a discussion of fractals in plant growth, in Will Benson, *Kingdom of Plants: A Journey through Their Evolution*, London: Collins, 2012.

p. 105 'Yet the remarkable passages': John Ruskin, 'Of Leaf Beauty', *Modern Painters*, Vol. IV, Part V, 1856.

Myths of Cultivation

10. The Celtic Bush: Hazel

p. 110 'short distances and definite places': Auden quotation from his poem 'In Praise of Limestone', 1948.

p. 113 'Oliver Rackham's more persuasive': Rackham's view of the 'Elm Decline': see *Woodlands*, op. cit.

p. 114 'One modern hazel worker': Hazel worker: Mark Powell, quoted in Richard Mabey, *Flora Britannica*, London: Sinclair Stevenson, 1996.

p. 118 'You could see the expansive': Irish hazel names: see Niall Mac Coitir, *Irish Trees: Myths, Legends and Folklore*, Wilton, Co. Cork: Collins Press, 2003.

p. 119 'The Burren's great biographer': Tim Robinson, *The Burren: A Map of the Uplands of North-west Clare, Eire*, Roundstone, Co. Galway: Folding Landscapes, 1977.

p. 120 'the greenwood that hangs': E. Charles Nelson and Wendy F. Walsh, *The Burren: A Companion to the World of Flowers*, Kilkenny: Boethius Press, 1991.

p. 121 'always [as] an emblem': Adrian Harris, quoted in Mabey, *Flora Britannica*, op. cit.

p. 121 'The story whose compact': Hazel mythology: see Coitir, *Irish Trees*, op. cit.

10. The Vegetable Lamb: Cotton

p. 122 'There growth there': Sir John Mandeville, quotation from an 1839 facsimile by J. O. Halliwell of a 1725 transcription of *The Voiage and Travaile of Sir John Maundevile*, London: Edward Lumley, 1839.

p. 123 'More than two centuries': Historical quotations taken from Henry Lee, *The Vegetable Lamb of Tartary: A Curious Fable of the Cotton Plant*, London: Sampson, Low & Co., 1887.

p.125 'A more plausible explanation': Sir Hans Sloane, Remark on a Letter of George Dampier, *Philosophical Transactions of the Royal Society*, Vol. 20, 1 January 1698, 52.

p. 126 'I have to express': Lee, *Vegetable Lamb*, op. cit.

p. 128 'To speake of the commodities': John Gerard, *The Herball, or Generall Historie of Plantes*, 1597, Thomas Johnson revised edn, 1636 (in a New York: Dover facsimile, 1975).

p. 128 'There are four members': History of cotton: see Edgar Anderson, *Plants, Man and Life*, London: Andrew Melrose, 1954, new edn Cambridge: Cambridge University Press, 1967.

p. 130 'In the heyday of spiritualism': Cotton in séances: see Marina Warner, *Phantasmagoria: Spirit Visions, Metaphors and Media into the Twenty-first Century*, Oxford: Oxford University Press, 2006.

p. 131 'But enduring myths': Richard Holloway, personal communication, and 'How to Film the Bible', *Daily Telegraph*, 5 April 2014.

p. 132 'The great biological essayist': Lewis Thomas, *The Lives of a Cell: Notes of a Biology Watcher*, New York: Viking Press, 1974.

11. The Staff of Life: Maize

p. 133 'The myths of corn's origin': Myths of maize origins: see John E. Staller, Robert H. Tykot Bruce F. Benz, *Histories of Maize in Mesoamerica. Multidisciplinary Approaches*, Walnut Creek, Calif.: Left Coast Press, 2010.

p. 133 'So too are the group of stories': Claude Lévi-Strauss, *From Honey to Ashes*, 1966, trans. John George and Doreen Weightman, London: Cape, 1973. For more maize origin myths see also his *The Raw and the Cooked*, 1964, trans. John George and Doreen Weightman, first UK edn London: Cape, 1970.

p. 136 'These early maize ears': Maize evolution: see Stoller et al., *Histories of Maize,* op. cit.; Anderson, *Plants, Man and Life*, op cit. See also Oliver Sacks, *Oaxaca Journal*, Washington, DC: National Geographic Society, 2001.

p. 137 'Forest gardening': Forest gardens: see Anderson, *Plants, Man and Life*, op. cit.; Gerardo Reichel-Dolmatoff**,** *The Forest Within: The Worldview of the Tukano Amazonian Indians*, Totnes: Themis, 1996.

p. 139 'During the mid twentieth century': Anderson, *Plants, Man and Life*, 1967, op. cit.

p. 141 'If this were a story': Evelyn Fox Keller, *A Feeling for the Organism: The Life and Work of Barbara McClintock*, San Francisco: Freeman, 1983.

12. The Panacea: Ginseng

p. 143 'That a genus': Ginseng history: see Barbara Griggs, *New Green Pharmacy: The History of Western Herbal Medicine,* 2nd edn, London: Vermilion, 1997; Andrew Dalby, 'Ginseng: Taming the Wild', *Proceedings of the Oxford Symposium on Food and Cookery 2004: Wild Food*, London; Prospect Books, 2006.

p. 145 'In the complex mythology': Tukano *uacú* ritual: see Reichel-Dolmatoff, *Forest Within*, op. cit.

p. 145 'In the redoubts': Appalachian indigenous medicine: see Anthony Cavender, *Folk Medicine in Southern Appalachia*, Chapel Hill, NC, London: University of North Carolina, 2003.

p. 148. 'Chimpanzees with infections': Self-medication in animals: see Hanson, *Triumph of Seeds*, op. cit.

p. 150 'tell the visionary': Reichel-Dolmatoff, *The Forest Within*, 1966, op. cit.

p. 150 'Amongst the Runa people': Eduardo Kohn, *How Forests Think: Toward an Anthropology Beyond the Human*, Berkeley: University of California Press, 2013.

p. 153 'The Cherokee believed': Cherokee beliefs: see Cavender, *Folk Medicine*, op cit.

p. 154 'It gives an uncommon': William Byrd quoted in Griggs, *New Green Pharmacy*, op cit.

p. 155 'Whole communities turned': Ginseng in US community life and economy: see Euell Gibbons, *Stalking the Healthful Herbs*, New York: McKay, 1966; Cavender, *Folk Medicine*, op. cit; Kristin Johannsen, *Ginseng Dreams. The Secret World of America's Most Valuable Plant*, Lexington: University Press of Kentucky, 2006.

13. The Vegetable Mudfish: Samphire

p. 160 'Diligent botanist': V. J. Chapman, 'Studies in Salt-marsh Ecology, Parts I–III', *Journal of Ecology*, Vol. 26, 1938, 144–79.

The Shock of the Real: Scientists and Romantics

p. 164 'It would be easy': Wilfrid Blunt, *The Compleat Naturalist: A Life of Linnaeus* , London: W. Collins, 1971.

p. 165 'I love to see the nightingale': John Clare's Journal in Margaret Grainger, ed., *The Natural History Prose Writings of John Clare*, Oxford: Clarendon, 1983; Marilyn Gaull, 'Clare and "the Dark System"', in *John Clare in Context*, ed. Hugh Haughton, Adam Phillips and Geoffrey Summerfield, Cambridge: Cambridge University Press, 1994.

p. 166 'In 1817, at a very lively': Haydon's 1817 party: Nicholas Roe, *John Keats: A New Life*, New Haven, Conn., London: Yale University Press, 2012.

14. Life versus Entropy: Newton's Apple

p. 167 'After dinner, the weather': William Stukeley, *Memoirs of Sir Isaac Newton's Life*, 1752.

p. 171 'At the start of the twenty-first century': Barrie E. Juniper and David J. Mabberley, *The Story of the Apple*, Portland, Ore.: Timber Press, 2006. p. 169'In the United States' Henry David Thoreau, *Wild Fruits: Thoreau's Rediscovered Last Manuscript*, ed. Bradley P. Dean, New York: W. W. Norton, 2000.

p. 171 'And in 1841': Story of 'Lane's Prince Albert': see Francesca Greenoak, *Forgotten Fruit: the English Orchard and Fruit Garden*, London: Deutsch, 1983; Joan Morgan and Alison Richards, *The New Book of Apples*, revised edn, London: Ebury, 2002.

p. 172 'The work eventually ran': *The Herefordshire Pomona* is available on a CD, published by the Marcher Apple Network, 2005.

15. Intimations of Photosynthesis: Mint and Cucumber

p. 176 'The nature of air obsessed': Jenny Uglow, *The Lunar Men; The Friends who Made the Future, 1730–1810*, London: Faber, 2002.

p. 176 'Priestley had been experimenting': Joseph Priestley, *Experiments and Observations on Different Kinds of Air*, London, 1775. For the full history of the discovery of photosynthesis, see Oliver Morton, *Eating the Sun: The Everyday Miracle of How Plants Power the Planet*, London: Fourth Estate, 2007.

p. 178 'In 1727 Stephen Hales': Stephen Hales, *Vegetable Staticks, or, an Account of Some Statical Experiments on the Sap in Vegetables*, London: W. & J. Innys, 1727.

p. 180 'I hope this will give': Benjamin Franklin's letter quoted in Priestley, *Experiments*, op. cit.

p. 183 'The atom of charcoal': Shirley Hibberd, *Brambles and Bay Leaves: Essays on the Homely and the Beautiful*, London, 1855.

p. 183 'It's a passage': Primo Levi, *The Periodic Table*, 1975.

16. The Challenge of Carnivorous Plants: The Tipitiwitchet

p. 184 'This letter quoted': E. Charles Nelson, *Aphrodite's Mousetrap: A Biography of Venus's Flytrap*, Boethius Press, 1990. I am indebted to this book for much of the detail of the flytrap's early history.

p. 185 'It put the usefulness': For an attack on argument by analogy see Philip C. Ritterbush, *Overtures to Biology: The Speculations of Eighteenth-Century Naturalists*, New Haven, Conn.: Yale University Press, 1964.

p. 189 'Excuse me if in vanity': William Logan's Ms letter is in the Aberdeen University Library.

p. 190 'This is a footnote': Erasmus Darwin, *The Botanic Garden*, London: J. Johnson, 1791. See also Molly Mahood's discussion of Erasmus in *The Poet as Botanist*, op. cit.

p. 191 'In December 1990': Daniel L. McKinley, quoted in Nelson, *Aphrodite's Mousetrap*, op. cit.

p. 192 'It's gratifying to see': Barry A. Rice, *Growing Carnivorous Plants*, Portland, Ore.: Timber Press, 2006.

p. 192 'In 1968, secondary school': Isle of Man teenagers: quoted in Mabey, *Flora Britannica*, op. cit.

p. 193 'During the summer of 1860': Charles Darwin, *Insectivorous Plants*, London, 1875.

p. 196 'He was thrilled': Research on the electrical activity in the Venus flytrap, in Daniel Chamovitz, *What a Plant Knows: A Field Guide to the Senses of Your Garden – and Beyond*, Richmond: One World, 2012.

17. Wordsworth's Daffodils

p. 198 'the world's most popular spring flower': Noel Kingsbury and Jo Whitworth, *The Remarkable Story of the World's Most Popular Spring Flower*, Portland, Ore.: Timber Press, 2013.

p. 200 'He wrote his famous': Ovid's *Metamorphoses*, translation by Rolfe Humphries, London: John Calder, 1955.

p. 201 'The Elizabethan botanical writer': Gerard, *Herball*, op. cit.

p. 201 'The first written': William Turner, *A New Herball*, London: Steven Mierdman, 1551.

p. 201 'And it had vernacular': Local daffodil names: Geoffrey Grigson, *The Englishman's Flora*, London: Phoenix House, 1955.

p. 201 'When the Belgian botanist': Charles de l'Écluse, quoted in Carolus Clusius, *Rariorum Plantarum Historia*, 1601.

p. 203 'Ted Hughes, in his superb poem': Ted Hughes, *Collected Poems*, ed. Paul Keegan, New York: Farrar, Straus & Giroux, 2003.

p. 204 'From Marcle Way': Lascelles Abercrombie poem, in Sean Street, *The Dymock Poets: Poetry, Place and Memory*, Bridgend: Seren Books, 1998.

p. 206 'We fancied that the lake': Dorothy Wordsworth, in Clark, ed., *Grasmere Journal*, op. cit.

p. 207 'Lucy Newlyn, in her brilliantly' Newlyn, *William and Dorothy Wordsworth*, op. cit.

18. On Being Pollinated: Keats's Forget-Me-Not

p. 209 'The Revd Richard Polwhele agreed': Revd Polwhele, quoted in Patricia Fara, *Sex, Botany and Empire: The Story of Carl Linnaeus and Joseph Banks*, Cambridge: Icon, 2003.

p. 210 'Nor can I find': Quotations from Keats's poems are from *John Keats: The Major Works*, ed. Elizabeth Cook, Oxford: Oxford University Press, 1990.

p. 211 'Thirty years previously': Joseph Kölreuter and Christian Sprengle: see Michael Proctor and Peter Yeo, *The Pollination of Flowers*, London: Collins, 1973.

p. 211 'The ever-imaginative': Erasmus Darwin, 'The Economy of Vegetation', *The Botanic Garden*, London: J. Johnson, 1791.

p. 212 '[He] experimented with': Patrick Blair, *Botanick Essays*, London: W. & J. Innys, 1720.

p. 212 'Now if the Bee': Arthur Dobbs, 'Concerning Bees and Their Method of Gathering Wax and Honey', *Philosophical Transactions of the Royal Society*, Vol. 46, 1750, 536–49.

p. 213 'Experiments involving scent': Insects' colour preferences: see Proctor and Yeo, *Pollination*, op. cit.

p. 215 'Did Keats suspect any': Keat's life: see Roe, *John Keats*, op. cit.; Robert Gittings, *John Keats*, London: Heinemann, 1968.

p. 215 'It has been an old Comparison': Letter to John Reynolds, quoted in Gittings, *John Keats*, op. cit.

p. 217 'Modern writers such as': For a robust defence of Romantic vitalism, see Denise Gigante, *Life: Organic Form and Romanticism*, New Haven, Conn., London: Yale University Press, 2009; David Rothenberg, *Survival of the Beautiful: Art, Science and Evolution*, London: Bloomsbury, 2011; Michael Pollan, *The Botany of Desire: A Plant's-eye View of the World*, London: Bloomsbury, 2002.

p. 218 '[These] are the real': John Ruskin, *Proserpina: Studies of Wayside Flowers*, 2 vols., Orpington: George Allen, 1875–6.

New Lands, New Visions

p. 220 'I reported my friend': David Nash exhibition at Royal Botanic Gardens, Kew, 2012.

p. 221 'One by one the plants': Mary- Louise Pratt, *Imperial Eyes: Travel Writing and Transculturation*, London: Routledge, 1992.

p. 221 'The natural-history artist': Kim Todd, *Chrysalis: Maria Sibylla Merian and the Secrets of Metamorphosis*, London: I. B. Tauris, 2007. See also the chapter on Merian in Natalie Zemon Davis, *Women in the Margins: Three Seventeenth-century Lives*, Cambridge, Mass.. London: Harvard University Press, 1995.

19. Jewels of the Desert: Francis Masson's Starfish and Birds of Paradise

p. 224 'For Britain's first official': For Masson's life, see a series of papers by M. C. Karsten in the *Journal of South African Botany*, Vols. 24–7, 1958–61.

p. 252 'This tract of country': Francis Masson, 'An Account of Three Journeys from the Cape Town into the Southern Parts of Africa', *Philosophical Transactions of the Royal Society*, Vol. 66, 1776, 319–52.

20. Growing Together: The East India Company's Fusion Art

p. 232 'India had been colonised': Company art: see Mildred Archer, 'India and Natural History: The Role of the East India Company, 1785–1858', *History Today*, Vol. 9, No. 11, Nov. 1959, 736–43; also her *Indian Painting for the British, 1770–1880*, Oxford: Oxford University Press, 1955.

p. 238 'The clear, blue, cloudless': William Hodges, *Travels in India*, London, 1793.

21. The Chiaroscuro: The Impressionists' Olive Trees

p. 240 'The olive tree is': Aldous Huxley, *The Olive Tree and Other Essays*, London: Chatto & Windus, 1936.

p. 240 'The first detailed painting': Stephen Harris, *The Magnificent Flora Graeca: How the Mediterranean Came to the English Garden* , Oxford: Bodleian Library, 2007.

p. 241 'He painted them': Van Gogh quotes from Steven Naifeh and Gregory White Smith, *Van Gogh: The Life*, London: Profile Books, 2011.

p. 241 'It sparkles like diamonds': Derek Fell, *Renoir's Garden*, London: F. Lincoln, 1991.

p. 244 'For a long while': For a scholarly demolition of the 'Ruined Landscape' myth see Rackham and Grove, *The Nature of Mediterranean Europe*, op. cit.

22. Local Distinctiveness: Cornfield Tulips and Horizontal Flax

p. 247 'We gathered the Ebony': John Sibthorp, quoted in Harris, *The Magnificent Flora Graeca*, op. cit.

p. 249 'Other Cretan endemics': John Fielding and Nicholas Turland, *Flowers of Crete*, London: Royal Botanic Gardens, Kew, 2005.

p. 250 'What is Samaria': Antonis Albertis, *The Gorge of Samaria and its Plants*, Heraklion, 1994

p. 251 'Their special botanical': Chasmophytes: see S. Snogerup, 'Evolutionary and Plant Geographical Aspects of Chasmophytic Communities', *Plant Life of South-West Asia*, ed. Peter H. Davis, Peter C. Harper and Ian C. Hedge, Edinburgh: Botanical Society of Edinburgh, 1971

p. 252 'John Fielding's phtographs': Fielding and Turland, *Flowers of Crete*, op. cit.

The Victorian Plant Theatre

p. 253 'Predicted by Newton's prism': Glass: see Isobel Armstrong. *Victorian Glassworlds. Glass Culture and the Imagination, 1830–1880*, Oxford: Oxford University Press, 2008.

p. 253 'Nathaniel Ward was a doctor': Nathaniel Ward: see David Elliston Allen, *The Victorian Fern Craze: A History of Pteridomana*, London: Hutchinson, 1969; Nicolette Scourse, *The Victorians and Their Flowers*, London: Croom Helm, 1983.

p. 257 'Gleaming rock crystals': Kate Colquhoun, *A Thing in Disguise: The Visionary Life of Joseph Paxton*, London: Fourth Estate, 2003.

23. 'Vegetable jewellery': The Fern Craze

p. 258 'The eruption of interest': Pteridomania: see Allen, *Victorian Fern Craze*, op cit.; Edward Newman, *A History of British Ferns and Allied Plants*, London: John van Voorst, 1840.

24. 'The Queen of Lilies': *Victoria amazonica*

p. 264 'The first European': Early discoveries of *Victoria amazonica*: Tomasz Anisko, *Victoria: The Seductress*, Kennett Square, Pa.: Longwood Gardens, 2013.

p. 265 'Robert (eventually Sir Robert)': Robert Schomburgk: see account in *Curtis's Botanical Magazine*, Vol. 73, 1846.

p. 267 'The war of spin': Dispute over name and early flowerings: see Wilfrid Blunt, *In for a Penny: A Prospect of Kew Gardens: Their Flora, Fauna and Falballas*, London: Hamilton, Tryon Gallery, 1978; and Anisko, *Victoria*, op. cit.

p. 269 'the lily enterprise': Armstrong, *Victorian Glassworlds*, op. cit.

p. 271 'The impression of the plant': Richard Spruce, *Notes of a Botanist on the Amazon & Andes*, London: Macmillan, 1908.

p. 272 'Several expensive folios': Walter Fitch, *Victoria Regia, or, Illustrations of the Royal Waterlily*, London, 1851.

p. 274 'You will observe that nature': Colquhuon, *A Thing in Disguise*, op. cit.

25. A Sarawakan Stinkbomb: The Titan Arum

p. 276 'The single flower': Beccari. An account of the discovery and flowering of the Titan arum is in *Curtis's Botanical Magazine*, Vol. 117, 1891.

p. 276 'When David Attenborough filmed': David Attenborough, interview in *Daily Telegraph*, 28 October 2008.

p. 279 'For twenty happy years': The story of Hardings Wood is in Richard Mabey, *Home Country*, London: Century, 1990, and *Beechcombings: The Narratives of Trees*, London: Chatto & Windus, 2007 (revised pbk edn *The Ash and the Beech: The Drama of Woodland Change*, London: Vintage Books, 2013).

p. 280 'In that first year': Richard Mabey, *Nature Cure*, London: Chatto & Windus, 2005.

p. 281 'And in a fen': J. E. Lousley:see Victor Samuel Summerhayes, *Wild Orchids of Britain*, 2nd edn, London: Collins, 1969.

p. 282 'Unsuprisingly, it was never seen': Rediscovery of Sander's slipper orchid: see Eric Hansen, *Orchid Fever: A Horticultural Tale of Love, Lust and Lunacy*, London: Methuen, 2000.

p. 285 'They resembled a clog': Joris-Karl Huysmans, *Against Nature,* 1884.

p. 285 'H. G. Wells wrote a brisk': H. G. Wells, 'The Flowering of the Strange Orchid', *The Works of H. G. Wells, Atlantic Edition*, Vol. I, *The Time Machine, The Wonderful Visit, and Other Stories*, London: T. Fisher Unwin.1924.

p. 285 'At the start of Raymond Chandler's': Raymond Chandler , *The Big Sleep*, 1939. For an overview of orchids in fiction: see Potter, *Seven Flowers*, op. cit.

p. 287 'The gardener Philp Miller': Philip Miller, *Gardeners' Dictionary,*, 4th edn, London, 1740.

p. 288 'It is symmetrically': Maurice Maeterlinck, *The Intelligence of Flowers*, 1907, translated by Philip Mosley, Albany, NY: State University of New York Press, 2008;. For a discussion of orchid mimicry, see H. Martin Schaefer and Graeme D. Ruxton, *Plant-Animal Communication*, Oxford: Oxford University Press, 2011.

p. 297 '"Nice lip," ventured': Hansen, *Orchid Fever*, op. cit.

p. 297 'No wonder John Ruskin': Ruskin, *Proserpina*, op. cit.

p. 298 'Their forms are beyond': Charlotte Mary Yonge, *The Herb of the Field*, London, 1853.

p. 300 'The story of its subsequent': Rediscovery of *Cattleya labiata vera,* in Tyler Whittle, *The Plant Hunters: 3450 Years of Searching for Green Treasure*, London: Heinemann, 1970. p. 303 'fresh air and the winds': Joseph Dalton Hooker, *Himalayan Journals*, 2 vols., London: John Murray, 1854.

p. 306 'in a state of some': Martin Sanford, *A Flora of Suffolk*, Ipswich: D. K. & M. N. Sandford, 2010.

p. 308 'In 2014 the journal': *Boquila*: see E. Gianoli and F. Carrusco-Urra, 'Leaf Mimicry in a Climbing Plant Protects against Herbivory', *Current Biology*, Vol. 24, no. 9, 5 May, 2014, 984–7.

27. The Butterfly Effect: The Moonflower

p. 311 'In 1972 the English artist': For Margaret Mee's life and journals, see Margaret Mee, *In Search of Flowers of the Amazon Forests*, ed.

Tony Morrison, Nonesuch Expeditions,, 1988, and Mee, *Flowers of the Brazilian Forest*, London: Tryon Gallery in association with George Rainbird,1968.

p. 321 'But in the mid 1990s': W. Barthlott, S. Porembski, M. Kluge, J. Hopke and L. Schmidt, 'Selenicereus wittii (Cactaceae: An Epiphyte Adapted to Amazonian Igapó Inundation Forests', *Plant Systematics and Evolution*, Vol. 206, No. ¼, 1997, 175–86.

p. 321 'They are typical': 'white-floral-image' see Roma Kaiser, *The Scent of Orchids: Olfactory and Chemical Investigations*, Elsevier, 1993.

28. The Canopy Cooperative: Air Plants and Bromeliads

p. 323 'One of the Amazonian': Mee, *Brazilian Forests*, op cit.; Margaret Mee and Lyman B. Smith, *The Bromeliads: Jewels of the Tropics*, South Brunswick: A. S. Barnes; London: Thomas Yoseloff, 1969. p. 324 'They build up symbioses': Rainforest symbioses: see Schaeffer and Ruxton, *Plant-Animal Communication*, op. cit.; Benson, *Kingdom of Plants*, op. cit.

p. 325 'One modern journalist': Charles C. Mann, '1491', *The Atlantic*, March 2002.

p. 325 'Mark W. Moffett, a Harvard': Mark W. Moffett, *The High Frontier: Exploring the Tropical Rainforest Canopy*, Cambridge, Mass., London: Harvard University Press, 1993.

p. 327 'Instead of climbing': Canopy rafts: see Adrian Bell, 'On the Roof of the Rainforest', *New Scientist*, 2 February 1991.

p. 327 'This is a useful corrective': Amazonia as a human artefact: see the arguments between Betty J. Meggers, *Amazonia: Man and Culture in a Counterfeit Paradise*, Chicago: Aldine Atherton, 1971, and Anna C. Roosevelt, *Moundbuilders of the Amazon*, San Diego, London: Academic Press, 1991.

29. Plant Intelligence: Mimosa

p. 331 'John Freke, an English John Freke, *Essay to Shew the Cause of Electricity*, London, 1746. p. 331 'George Bell, in 1776': George Bell, quoted in Ritterbush, *Overtures to Biology*, op. cit., which contains an overview (rather hostile) of the use of analogy in eighteenth-century botany.

p. 331 'So sinks or rises': Erasmus Darwin, *Loves of Plants*, op. cit.

p. 332 'When the sequence of events': Mechanism of mimosa's folding: see Dov Koller, *The Restless Plant*, Cambridge, Mass., London: Harvard University Press, 2011; Chamovitz, *What a Plant Knows*, op cit. Monica Gagliano's mimosa experiment, see Michael Pollan, 'The Intelligent Plant', *The New Yorker*, 23 December 2013.

p. 335 'chemical communication between plants': Plant-plant communication:
see Stefano Mancuso and Alessandra Viola, *Verde Brilliante:
Sensibilata e intelligenza del mondo vegetale*, Florence, 2013; Schaefer
and Ruxton, *Plant-Animal Communication*, op. cit. mopane trees; Colin
Tudge, *The Secret Life of Trees: How They Live and Why They Matter*,
London: Allen Lane, 2005.

p. 335'of the University': Suzanne Simard's research outlined in Pollan, 'The
Intelligent Plant', op. cit.

p. 336 'But sound communication': E.g. Monica Gagliano, 'Green Symphonies:A
Call for Studies on Acoustic Communication in Plants', *Behavioral
Ecology,* Vol. 24, No. 4, 2013, 789–96; M. Gagliano and M. Renton, 'Love
Thy Neighbour: Facilitation through an Alternative Signalling Modality in
Plants', *BMC Ecology*, Vol. 13, May 2013.

p. 337'The idea that': Daniel Dennett quoted in Pollan, 'The Intelligent Plant',
op cit. For discussions of vegetal intelligence, see 'The Aware Plant' in
Chamovitz, *What a Plant Knows*, op. cit., and Michael Marder, *Plant-
Thinking. A Philosophy of Vegetal Life*, New York, Chichester; Columbia
University Press, 2013.

Epilogue

p. 339 'Forty years after': A new edition of *Beechcombings* has been published
as *The Ash and the Beech: The Drama of Woodland Change*, London:
Vintage Books, 2013.

List of Illustrations

the Bancroft Library, University of California, Berkeley (BANC PIC 1963.002:0381--B)

p. 81 Bristlecone, Inyo National Forest, California. Photo: National Geographic/Getty Images

p. 84 Preserved frond of the cycad Nilssonia kendalli, from Yorkshire, England. Photo: © The Trustees of the Natural History Museum, London

p. 94 Boars eating acorns under an oak tree, with a shepherd. Illustration from *Tacuinum sanitatis*, fourteenth or fifteenth century, a Latin translation of the *Taqwim al-sihha* (*Maintenance of Health*), an eleventh-century Arab medical treatise by Ibn Butlan of Baghdad. Photo: akg-images/Album/Oronoz

p. 100 John Crome, *The Poringland Oak*, c. 1818–20. Tate Gallery, London. Photo: © Tate, London

p. 103 The Green Man of Bamberg, thirteenth-century, Bamberg Cathedral, Germany. Photo: Bildarchiv Monheim/akg-images

p. 105 'From little acorns …' snow sculpture by Lisa Lindqvist and Kate Munro in Kiruna, Sweden, 2014

p. 116 Early Purple Orchid, the Burren, Ireland. Photo: © Carsten Krieger

p. 124 The Vegetable Lamb (*Agnus Sciithicus*), illustration by Matthäus Merian the Younger, from Jan Jonston, *Dendrographias*, 1662. Photo: (c) RHS, Lindley Library

p. 127 Cotton plant. Photo: Snapwire/Alamy

p. 135 Dried corncobs, Tarma province, Junín region, Peru. Photo: Aurélia Frey/akg-images

p. 138 John White, *Bird's eye view of the Indian village of Secoton (Secota), c.* 1570–80. British Museum, London. Photo: Bridgeman Images

p. 152 Shamanic painting of the mountain deity, Sansin, holding a root of wild ginseng. Korean school, undated. Photo: akg-images/Mark De Fraeye

p. 161 Glasswort (*Salicornia europaea*), a type of samphire, growing on estuary mud, The Wash, England. Photo: Michael Clark/FLPA/Rex

p. 173 Plate VIII from *The Herefordshire Pomona*, 1878–1885, ed. Robert Hogg & Henry Graves Bull, illustrated by Alice Ellis and Edith Bull. Reproduced by kind permission of the Folio Society

p. 179 Illustration by Milo Winter from Jonathan Swift, *Gulliver's Travels*, 1912

p. 181 Electron micrograph of a mint leaf. Photo: Annie Cavanagh/Wellcome Images

p. 187 Venus flytrap (*Dionaea muscipula Ellis*), from L. van Houtte, *Flore des serres et des jardin de l'Europe*, vol. 3, 1847. Photo: courtesy Missouri Botanical Garden, St Louis (www.botanicus.org)

p. 194 Fly trapped on Round-leaved Sundew, Bankhead Moss, Fife, Scotland. Photo: © Ben Dolphin

p. 205 Walter Crane, *La Primavera* (detail), 1883. Private Collection. Photo: Roy Miles Fine Paintings/Bridgeman Images

p. 214 Forget-me-not. Photo: Getty Images

p. 223 Maria Sibylla Merian, *Branch of guava tree with leafcutter ants, army ants, pink-toed tarantulas, huntsman spiders, and ruby topaz hummingbird*, c. 1701–5. Royal Collection Trust © Her Majesty Queen Elizabeth II, 2015. Photo: Bridgeman Images

p. 227 *Stapelia Gordoni*, illustration by Francis Masson from *Stapeliae novae, or, A collection of several new species of that genus, discovered in the interior parts of Africa*, 1796. Photo: Special Collections, Edinburgh University Library

p. 234 *Caesalpinia sappan L.* Hand-painted copy of an illustration by a Company artist commissioned by William Roxburgh for *Plants of the coast of Coromandel*, vol. 1, 1795. Photo: Kew Images/William Roxburgh Collection. Copyright © RBGKew

p. 237 *Lagerstroemia speciosa (L.) Pers.* [as *Lagerstroemia reginae Roxb.*] Hand-painted copy of an illustration by a Company artist commissioned by William Roxburgh for *Plants of the coast of Coromandel*, vol. 1, 1795. Photo: Kew Images/William Roxburgh Collection. Copyright © RBGKew

p. 242 Vincent van Gogh, *The Olive Grove*, 1889. Rijksmuseum Kröller-Müller, Otterlo. Photo: De Agostini Picture Library/Bridgeman Images

p. 248 *Tulipa doerfleri*, Crete. Photo: Paul Harcourt Davies/Science Photo Library

p. 260 A Wardian case in a bay window, from *Cassell's Home Guide*, nineteenth century. Private Collection. Photo: Look and Learn/Bridgeman Images

p. 262 Nicholas Chevalier, design for a dress for Lady Barkly, the wife of the Governor of Victoria, 1860. Photo: Courtesy of the National Library of Australia, Canberra (NK599)

p. 266 Walter Fitch, detail drawing of the flower and leaf structure of the *Victoria amazonica* growing at Syon House, illustration from William Hooker, *Victoria regia; or Illustrations of the Royal Water-Lily*, 1851, plate 4.

p. 273 A young man holding an enormous Victoria Regia pad, 1936. Photo: Hulton-Deutsch Collection/Corbis

p. 277 Timelapse photo of growth of Titan, composite photograph from a webcam. Photo: Botanischer Garten der Universität Basel

p. 283 *Cypripedium Sanderianum*, illustration by Frederick Sander from *Reichenbachia I*, plate 03, 1888. Photo: courtesy Missouri Botanical Garden, St Louis (www.botanicus.org)

LIST OF ILLUSTRATIONS

p. 285 Martin Johnson Heade, *Cattleya Orchid and Three Hummingbirds*, 1871. National Gallery of Art, Washington, D.C. Accession N0.1982.73.1 Gift of The Morris and Gwendolyn Cafritz Foundation

p. 301 *Native Dinner-Time*, illustration from Albert Millican, *Travels and adventures of an orchid hunter. An account of canoe and camp life in Colombia, while collecting orchids in the northern Andes*, 1891. Photo: Brown University Library, Providence, R.I.

p. 313 Margaret Mee, pictured during her search for the moonflower, Rio Negro, Brazil, 1988 © Tony Morrison/Nonesuch Expeditions/South American Pictures

p. 319 Margaret Mee, *Selenicereus wittii (Cactaceae), The Amazon Moonflower*, 1988. Photo: © Tony Morrison/Nonesuch Expeditions/South American Pictures

p. 326 Bromeliads growing on a tree, Molinuco nature reserve, Ecuador. Photo: Sinclair Stammers/Science Photo Library

p. 330 George Romney, *Emma Hamilton in an attitude towards a mimosa plant, causing it to demonstrate sensibility*, 1789. Photo: Wellcome Library, London

p. 341 Frithsden Beeches, Ashridge Woods, Hertfordshire. Photo: Robert Stainforth/Alamy

Section illustrations

p. 10 Red, Blue, and White Lotus of the Hindostan. Photo: Getty Images/ Buyenlarge

p. 41 *Agathis australis* (antique botanical engraving). Photo: istockphoto/ nicoolay

p. 108 Apple Cultivar or *Calville blanc d'hiver* (Redoute Botanical Illustrations). Photo: istockphoto/nicoolay

p. 164 Antique illustration of *Lamium album* (white nettle). Photo: istockphoto/ilbusca

p. 220 European Fan Palm (*Chamaerops humilis*) Arecaceae by Angela Rossi Bottione, watercolor, 1806–1812. Photo: Getty Images/Dea/G. Cigolini

p. 253 Belgium, The Victoria Regia greenhouse of the Van Houtte Nursery in Gent, engraving. Photo: Getty Images/Dea/G. Dagli Orti

p. 308 Hortulus botanicus pictus sive collectio plantarum. Volume I, plate: Acacia (*Acacia Indica*), Giovanni Battista Morandi, 1748. Photo: Getty Images/Dea/G. Cigolini/Veneranda Biblioteca Ambrosiana

While every effort has been made to contact copyright-holders of illustrations, the author and publishers would be grateful for information about any illustrations where they have been unable to trace them, and would be glad to make amendments in further editions.

Acknowledgements

It's always difficult to judge how deep a book's roots go, so if there are any colleagues who recognise their influence in these pages and are not acknowledged below, let me apologise and thank them now. To those involved during the three years I was working directly on *Cabaret*, for reading and commenting on sections, for advice and support, for pointing out sources, stories, books, for company on walks and more ambitious explorations and for simple support – warm thanks to Bill Adams, Ian Collins, Jill Cook at the British Museum, Jon Cook, Francesca Greenoak, Jay Griffiths, Robin and Rachel Hamilton, Caspar Henderson, Richard Holmes, Tony Hopkins, the staff of the Hortus Botanicus in Amsterdam, Kathleen Jamie, the Linnaean Society of London, Robert Macfarlane, Lucy Newlyn, Peter Newmark, Norwich Cathedral Library, the staff of the Orto Botanico in Padua, Philip Oswald, Martin Sanford of the Suffolk Biological Records Centre, Martha, Reuben and Kit Shawyer, the late Richard Simon, and Christopher Woodward of the Museum of Garden History. And a special debt of gratitude to my dear friends and botanical mentors, Bob Gibbons and Libby Ingalls, for their wisdom and company on many botanising trips at home and abroad, and for sharing their experiences of places I have never visited.

Of the many sacks of books one reads in the course of research, I will single out one that has been especially inspirational: Molly Mahood's *The Poet as Botanist*. *The Cabaret* could be seen as a response, maybe subtitled *The Botanist as Poet*, or even *The Plant as Poet*.

I am deeply grateful to Dame Fiona Reynolds, Master of Emmanuel College, Cambridge, and the college fellows, for the award of a Derek Brewer Visiting Fellowship in 2014, which gave me the freedom of weeks of uninterrupted reading, and the chance to use the Cambridge University Botanic Gardens as my home patch. A particular bow is due to Emmanuel's Senior Arboreal Fellow, the Great Plane under whose prehensile boughs I spent many meditative and productive hours. In the Cambridge Botanic Gardens special thanks

ACKNOWLEDGEMENTS

to the glasshouse curator, Alex Summers and his staff, who gave freely of their time and knowledge. The cooperation of the Royal Botanic Gardens, Kew was also invaluable, and I thank the Director, Richard Deverell, for facilitating this, and Richard Barley, Director Horticulture, and the irrepressibly friendly and enthusiastic staff of the herbarium, archives In the engine room, great thanks to Andrew Franklin, my publisher at Profile, for commissioning the book and for his unwavering support even at times when the project must have seemed as unruly as a bramble patch. To Penny Daniel, my editor, for her patience and understanding during the book's progress, and my all too frequent lapses of faith in it, and for her diligent and sensitive editing of the final text. To Trevor Horwood for, as usual, forensically precise copy-editing and, thank goodness, for challenging me when I went too far out on a limb. To Cecilia Mackay for her brilliant work in finding and helping choose the images. To Douglas Matthews for his scholarly but accessible index. On the home front, my partner Polly was a sharp-eyed and enthusiastic companion on field trips, and warm support during the inevitable emotional ups and downs of writing a long book. She knows the score by now, but I'm well aware that doesn't make it any easier for her.

Finally more thanks than I can properly express to my agent Vivien Green. This book evolved from an original idea by her, and during its progress she acted as critic, therapist, arbitrator, applauder and purveyor of tough love with all the affection and wisdom she has shown over three decades. This book is dedicated to her.

Index

INDEX

INDEX